SYMPHONY OF THE EARTH

N.V. VAN DE GARDE & CO'S DRUKKERIJ, ZALTBOMMEL

# SYMPHONY

## OF THE

## EARTH

BY

### J. H. F. UMBGROVE

D. SC.; HON. F.R.S.E.
MEMB. R. NETHERL. AC. SC.
HON. MEMB. N.Y. AC. SC.
PROFESSOR OF GEOLOGY AT DELFT
HOLLAND

*With 10 plates and 123 textfigures*

SPRINGER-SCIENCE+BUSINESS MEDIA, B.V.

1950

ISBN 978-94-011-8668-1    ISBN 978-94-011-9463-1 (eBook)
DOI 10.1007/978-94-011-9463-1

*To*
*REGINALD ALDWORTH DALY*
*a constructive expression*
*of the author's high respect*
*for his masterly and inspiring leadership*
*in the development of*
*earth science*

# CONTENTS

# PREFACE

On several occasions I have been asked whether topics of earth science could be explained in more or less conventional prose. Of course this can be done. I hope chapter I meets with the most extreme wishes in this respect. I even venture to believe that chapters II, IV, VI and VII can also be digested without special strain. Though chapter V, and parts of chapter III may require a special effort of concentration on the part of the reader, there should be nothing in this book that cannot be understood by an intelligent person who is willing to try. Sometimes, however, the use of technical terms cannot well be avoided without affecting the meaning and accuracy of the ideas. However, I have attempted to restrict the use of these terms to a minimum. Moreover most scientific expressions and terms will be explained when used for the first time and when occurring again one will find a reference to these places in the index. Finally a list of the names of geological formations in their proper sequence is made available for consultation in table I, at the end of the Preface (p. XII).

Each chapter constitutes a subject in itself. However, as in the author's "The Pulse of the Earth" the reader will find that the different topics demonstrate the deeper correlation that links together a variety of terrestrial processes, which — at first sight — appear to be unconnected.

Each chapter is adapted from one or more addresses and lectures delivered on various occasions and in different countries. (Full reference to these, as well as to the reading literature, is made at the end of each chapter). Accordingly the text closely follows the spoken word.

Chapter I. *Symphonia terrestris*. Time and again the different themes of earth science will appear to be intimately interwoven parts of the great and fascinating symphony of the earth. This is what the first chapter tries to make clear in a general review, a sketch of the broad outlines, a score of the terrestrial symphony.

Some special topics come up for a fuller discussion in the ensuing chapters.

Chapter II. *A country below sea-level* is a sort of fascinating detective story. Hardly anything could be more dull than the starting point, viz. a few layers of clay and peat in the western provinces of the Netherlands. However, tracing their origin means unravelling the intricate interplay of movements of land- and sea-level. It will be demonstrated that the only means to disentangle this problem is furnished by data from oceanic and continental coasts bordering the seven seas.

Chapter III. *Across the Alps* might bear as undertitle: experimenting with a tablecloth. The rumpled tablecloth is represented by the Jura Mountains, the table by their rigid basement and the two are separated by a lubricating layer. By what processes was the tablecloth pushed into folds? To obtain a more or less satisfactory answer the whole Alpine belt has to be crossed from north to south, and a few excursions have to be made sideways. Some of these trips are far from easy going, which in a book like this means difficult reading. At the end, however, we return to the starting point and try to answer the first question.

Chapter IV. *Deep furrows on the continents and in the deep-sea.* The foregoing chapter touches on processes acting deep under the Alpine scenery. In order to study what processes probably take place in the deep realms beneath a folded mountain belt no region is more illuminating than the island-arcs of the East and West Indies. It is not the aim of this very short chapter to trace their geological history in its various aspects. The principal point which will be considered briefly is the origin of the deep-sea troughs accompanying the rather recent strip of mountain building in the East Indies. Moreover, structural zones of the western part of the East Indies have their counterpart in similar structures on the Asiatic continent. On the other hand tracing the continental structures back towards the East Indies furnishes a means of elucidating certain aspects of the submarine furrows.

Chapter V. *The root of a mountain chain.* After an excursion towards the deep-sea furrows of the East Indies and their counterparts in Burma, it will be less difficult to follow the author when he tries to plumb the depths beneath the Alps. Apart from results obtained in the two previous chapters, geophysical data will form the main guiding lines in exploring these deep-seated realms. This chapter forms as it were the completion of the picture developed in chapter III which was concerned with the more superficial features of the same area.

Chapter VI. *A trip on a volcano.* Apart from the mountain chains, which were considered in the three previous chapters, volcanoes and their associated phenomena offer another means of plumbing the depth of the earth. Starting from a trip on Mount Vesuvius several questions of general interest will be discussed. This means travelling mentally also in several other volcanic districts — Scotland, Mount Etna, New Hampshire, Sumatra, the Pacific, etc. — and, at the end, returning to the starting point.

Chapter VII. *Life and its evolution.* The last chapter is concerned with one of the problems mentioned cursorily on the last page of chapter I. It presents some meditations on the characteristics and evolution of life during the long eras of earth history and on the influence of the physical environment in the broad sense of the word. Some of the modern views on the problems of life and matter can only be expressed in non-pictorial concepts. But what then is the standard of valuation in these theories? In a short final section it is attempted to give a concise answer to these absorbing questions; in the meantime something is left to the reader to think over for himself.

Professor Ph. H. Kuenen has given generous assistance by reading the manuscript and Professor Arthur Holmes went through the page proofs. I owe much to their constructive criticism and valuable suggestions.

Seven of the figures have appeared already in "The Pulse of the Earth" (viz. figs. 3, 7, 9, 11, 12, 41, 67) and twenty five other text-illustrations have been taken from "Leven en Materie" (viz. figs 17, 94–99, 101–112 and 115–120). They were executed by Mr. C. van Werkhoven and Dr. R. de Wit respectively.

To Mr. J. Vuyk I owe the final drafting of the other 91 text illustrations as well as of Plates III, IV, V, VI and VII.

Plates III A, IV B, V A, V C, and IX A are reproduced from photographs which were on sale as picture-postcards. The other reproductions of landscapes are after photographs by the present author.

It is a pleasure to record my cordial thanks to the Secretary of the Royal Netherlands Academy of Sciences at Amsterdam for supplying the blocks of figures 20, 23, 28, 30, 32, 33, 42, 46–50, 57, 58, 61–65 and Plate III.

I also desire to express my thanks to the Council of the Dutch Society of Sciences (*Hollandsche Maatschappij van Wetenschappen*) at Haarlem for their permission to publish chapter I which, in a non-illustrated Dutch version, appeared previously in the series *Haarlemse Voordrachten* (No. VII, De Erven F. Bohn, N.V. Haarlem, 1948).

To Mr. W. Nijhoff and his staff I am very grateful for the care and courtesy which they have given to all the details involved in the production of this book.

## TABLE I
### Geological formations with their approximate duration in millions of years [1]).

| Eras | Periods and systems | | Approximate ages in millions of years |
|---|---|---|---|
| Cenozoic | Pleistocene . . . . . . . . . . . . | | 1 |
| | Tertiary . . . . . . . . | Pliocene | 1–12 |
| | | Miocene | 12–26 |
| | | Oligocene | 26–38 |
| | | Eocene | 38–58 |
| Mesozoic | Cretaceous. . . . . . . . . . . . . | | 58–127 |
| | Jurassic . . . . . . . . . . . . . . | | 127–152 |
| | Triassic . . . . . . . . . . . . . . | | 152–182 |
| Paleozoic | Permian . . . . . . . . . . . . | | 182–203 |
| | Carboniferous . . . . . . . . . . . . | | 203–255 |
| | Devonian . . . . . . . . . . . . | | 255–313 |
| | Silurian . . . . . . . . . . . . . . | | 313–350 |
| | Ordovician. . . . . . . . . . . . | | 350–430 |
| | Cambrian . . . . . . . . . . . . | | 430–510 |
| Pre-Cambrian (known rocks) . . . . . . . . . . | | | 510–2000 |
| Origin of the Earth . . . . . . . . . . . . . . | | | about 3300 |

| | | | |
|---|---|---|---|
| Names of Tertiary Stages and substages used in Chapter III | Pliocene | Pontian | |
| | Miocene | Tortonian | |
| | | Helvetian | |
| | | Burdigalian | |
| | | Aquitanian | |
| | Oligocene | Upper (Stampian, Rupelian) | |
| | | Lower | |
| | Eocene | | |

[1]) Time-scale according to recent data by A. Holmes. The scale of the figures in the text are based on an older time-scale.

CHAPTER I

# SYMPHONIA TERRESTRIS

*Introduzione e Fuga.*

No longer than a few decades ago, a young man announcing that he was going to study Geology might be answered by an eulogy on the clerical career and the study of Theology.

At present such a confusion of terms is not very likely to occur. In our country, too, the word "geology" sounds familiar and everybody is aware that geologists are interested in the structure and history of this our planet. But mostly the general interest does not go much farther than coal or oil, gold or precious stones.

Only few realize that the realm of earth science extends from the infinitely remote ages and depths of the universe to the origin and meaning of all organisms including the inmost depths of ourselves. Studying problems of earth science and examining their relation to other phenomena the route inevitably leads to these two extreme poles of human thought. It does not matter where we start, because all phenomena appear to be interrelated and each portion of the universe will come up at a certain moment to play its own part. We may start with any given landscape, in our own country or in the East Indies, in the Pacific or in Canada, in Italy or in Russia; we shall always end at these two extreme poles of thought.

To-day I should like to start by viewing a bit of scenery that most of you, if not all, have seen personally, viz., the lake of Lucerne, especially its north-eastern corner, where we see the little town of Brunnen, and higher up on the mountains the Mythen rocks.

This landscape may serve as the starting point for a short and summarizing exposition of the different motives which are interwoven in so masterly a way to create the great *Symphony of the Earth*. In rough outlines I shall try to show how the score of the terrestrial symphony is made up.

We start then at the Lake of Lucerne. Fig. 1 (A) shows Brunnen and the Mythen. Like everybody else a geologist enjoys the beautiful scenery.

Umbgrove

But then, — after a short while — a feeling of unrest gets hold of him.
What is the meaning of the huge stone-blocks rising from the gentle
slope below them like two threatening watchmen on a broad pedestal?
Do these tooth-shaped rock masses rest on the lower mountains or do
they pierce through them from deep below? Asking the question means
to grasp hammer and clinometer-compass, and the exploration, *mente
et malleo*, has begun.

After some time the geologist returns; his rucksack is heavily loaded
with a great number of rock fragments, and with him he brings many
notes, photographs and maps, which are painted in strongly contrasting
colours.

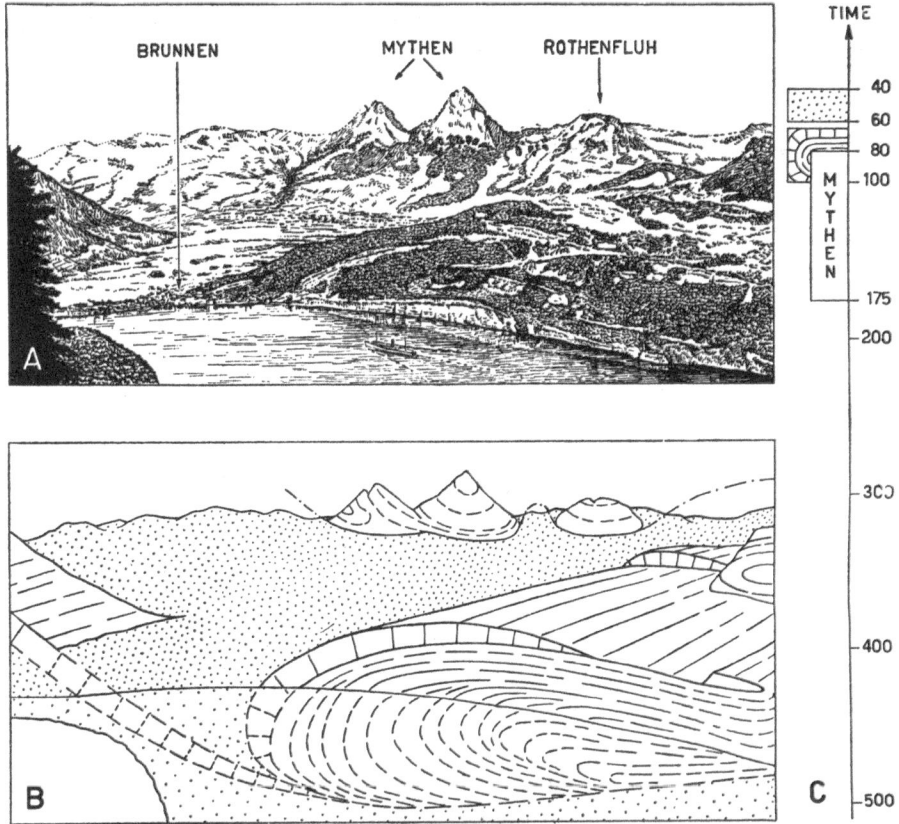

Fig. 1.  (A) Landscape along the northeastern coast of the Lake of Lucerne showing Brunnen
           and the Mythen.
        (B) The same landscape in geological notations.
        (C) Geological time-scale, in millions of years, showing the different ages of the rocks
           that make up the landscape near Brunnen.

Perhaps one is inclined to think the colours represent something like different types of rocks. That may be true but mostly each colour represents a special interval of time in the remote past. Fig. 1 (B) is such a geological sketch [1]). It shows that the Mythen do not pierce through the lower mountains but rest upon them. Moreover the colours demonstrate that the limestones of which the Mythen consist were formed in a time which is much more remote than the time when the underlying strata of the lower mountains originated. In order to elucidate the meaning of this remarkable result the line of fig. 1 (C) represents about the last sixth of geological time, i.e. 1/6 of the total span of time since the earth came into being. The lower part starts at a time about 500 million years ago while we are living so to speak at the zero top. The period of origin of the Mythen limestone is approximately indicated by the white column between the numerals 175 and 80; the period of the dotted terrain below the Mythen by the similar notation between 60 and 40.

Fig. 1 (A) and 1 (B) show yet another remarkable feature of the landscape. South of Brunnen a sequence of greyish limestones dip northward and disappear below the level of the lake, only to reappear to the north of Brunnen. It appears that these rocks are again older than the "dotted" layers below which they occur. Moreover, the "block limestones" were not called into being in the same surroundings as the calcareous rocks of the Mythen, most of which are still older as may be seen by comparing the time-scale, fig. 1 (C).

On further examination, covering a much wider area of the Swiss Alps, we shall find the solution of the remarkable time relations of the rock sequences in the Brunnen district. Fig. 2, a block-diagram of the High Calcareous Alps, shows several huge complexes of strata which rest on each other like a pile of blankets. They are called *nappes* and it appears that these nappes were moved from south to north. The limestones south of Brunnen belong to a pile of nappes which were moved northward from behind the Aar- and Gothard massifs. The Mythen, however, are mere fragments of a nappe which must have arrived from an area still farther south.

Evidently, our tentative sketch of the situation is only a very fragmentary representation resulting from many years of hard work and skilful examinations by several most able and enthusiastic geologists. In the meantime

---

[1]) For the sake of simplicity different black-white notations are used in the illustrations of this chapter. Dotted represents yellow, horizontal lines: blue, a block notation or diagonal lines descending from left to right: green.

the scenery near Brunnen leads us to unravel the structure of the Alps which from a geological point of view, belong to one of the best known mountain-chains. They can be followed eastward, through the Carpathians towards Asia Minor and via the Himalayas towards the East Indies.

However, a geologist is not yet content when considering a block-diagram like fig. 2. Although it gives a clear insight into the structure of a part

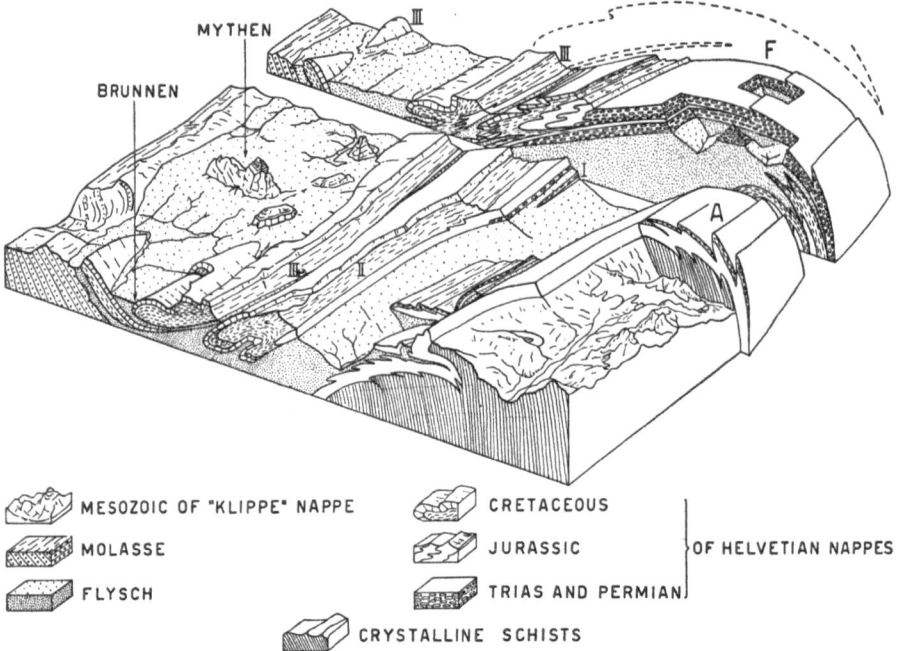

Fig. 2. Structural diagram of the surroundings of Brunnen and the Mythen (After P. Arbenz).

of Switzerland it is only a motionless representation of the present situation. His restless brains want to decipher the structural history of the whole area. He wants to see the past as a historical succession of phenomena and events up to the present day like the successive images of a cinematograph film, and if possible he wants to form an idea of the future stages of that particular region.

The first picture of the film shows a quite different scenery. The dominating feature is an elongated downwarp of the earth's surface forming a trough, more or less like block I in fig. 3. The bottom of the trough subsides gradually, but at the same time the furrow is filled up with deposits from the adjoining areas. This twofold process of an extremely

slow downward movement of the bottom combined with the accumulation of sediment may give origin to an abnormally thick sequence of
strata. Geologists call such an area a *geosyncline*. A thickness of several
miles, up to 10 miles and more is found in several places. It is not

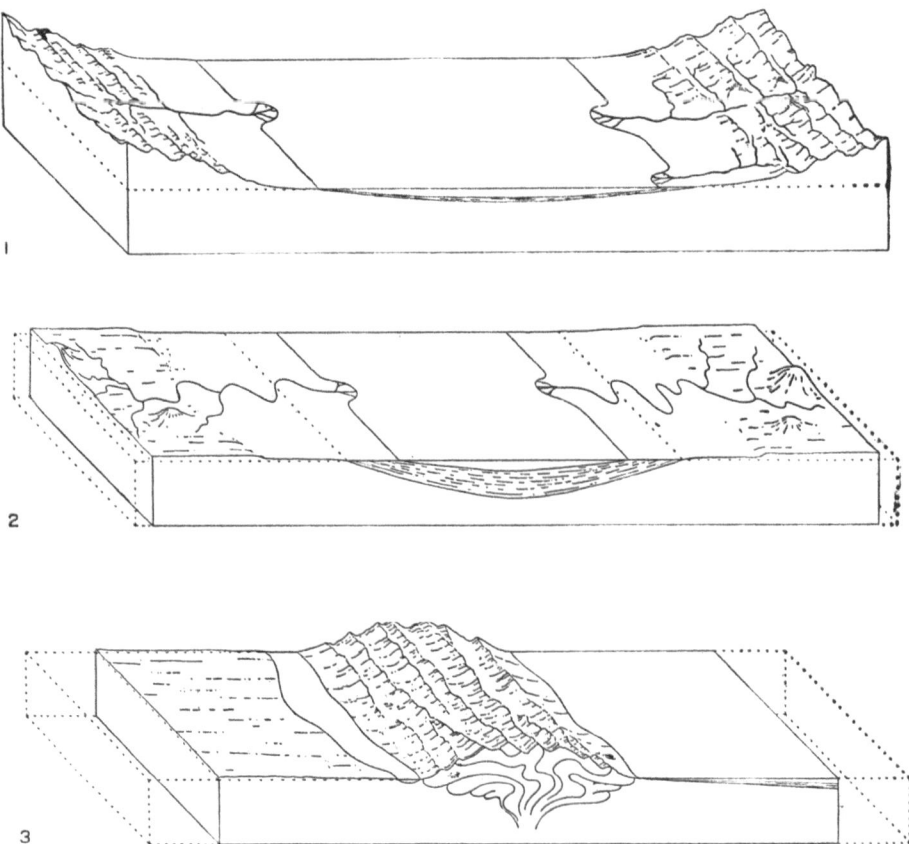

Fig. 3. Birth of a folded mountain-chain.
(1) and (2) Origin of a subsiding trough, gradually filled up with deposits from the adjoining areas.
(3) Compression of the trough, folding and crumpling of its contents, subsequent rising up of
the folded zone

difficult to imagine what must happen during a subsequent period of
strong compression of the earth's crust. The contents of the furrow will
become crumpled and folded. Some of the large folds will protrude over
the border of the trough and slip onto the adjacent "foreland". As a matter
of fact, the landscape near Brunnen represents such a border region. In a
more advanced stage (fig. 3 block 3) the whole zone will tend to rise;
for, the thick pile of light sedimentary rocks will tend to re-establish

equilibrium as soon as it gets the opportunity to do so, just as a submerged wooden log pressed down in water will rise upward as soon as the force which pressed it down is taken away. Consequently, from the elongated belt emerges a mountain range which very slowly rises upward though with unremitting persistence, frequently to a height of several thousands of metres.

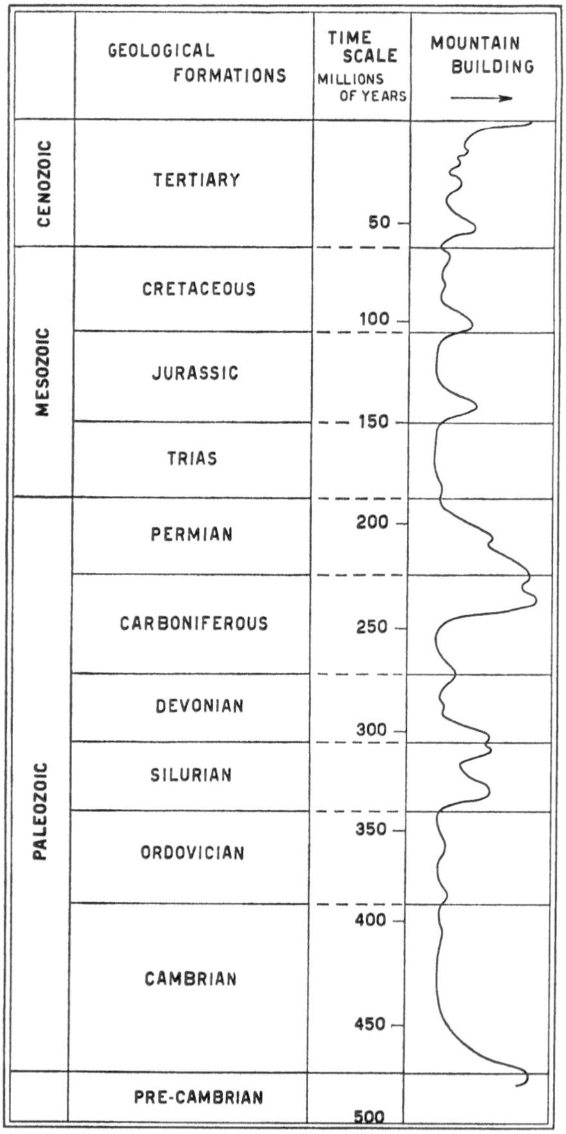

| GEOLOGICAL FORMATIONS | | TIME SCALE MILLIONS OF YEARS | MOUNTAIN BUILDING ⟶ |
|---|---|---|---|
| CENOZOIC | TERTIARY | 50 | |
| MESOZOIC | CRETACEOUS | 100 | |
| | JURASSIC | 150 | |
| | TRIAS | | |
| PALEOZOIC | PERMIAN | 200 | |
| | CARBONIFEROUS | 250 | |
| | DEVONIAN | 300 | |
| | SILURIAN | | |
| | ORDOVICIAN | 350 | |
| | CAMBRIAN | 400 / 450 | |
| | PRE-CAMBRIAN | 500 | |

Fig. 4. Graphic representation of three major periods of mountain-building known from the last 500 million years. The smaller tops of the curve indicate minor epochs of mountain-building.

External forces, i.e. the action of the atmosphere, running water, frost and ice embark upon their destructive action as soon as the folded chain emerges. A picture of the future shows demolition and denudation, acting slowly but irresistably until the monotonous stage of a peneplain will be completed.

The contents of the elongated furrow, from which the Alps were born, were folded in several stages. The rising movement which followed upon the last stage of compression and which gave origin to the majestic mountain-chain of our days, started about one million years ago. A glance at the time-scale (fig. 4) shows that this means hardly one tick back of the geological clock.

Much older are the remnants of an enormous folded zone, which can be studied in the Ardennes, the Eifel, the Harz mountains, and several other regions in Europe as well as in Asia and North America.

These structures, the denuded fragments of a mountain-chain probably surpassing the present Alps in majestic grandeur, originated about 250 million years ago.

Again about 250 million years earlier, certain zones of the earth were similarly subjected to strong compression and to subsequent mountain making. Tracing earth history still further backwards one finds similar processes in at least five further epochs separated from each other by analogous intervals of time.

The three last periods of strong folding have been noted on the accompanying graph (fig. 4) which also shows the names of the principal geological formations. The cause of the rhythmic recurrence of these phenomena is one of the fundamental problems of earth science, which cannot be attacked without entering the domain of physics and chemistry.

The rhythm displayed by folded structures is not the only manifestation of deep-seated processes. Numerous basin-shaped depressions originated in the earth's crust. Fig. 5 shows an example. It offers a simplified representation of the geological map of a part of France. A coloured map of the same area shows a few concentric blue and green rings surrounding a yellow coloured central area. (In fig. 5 they are represented respectively by horizontal and diagonal lines and a dotted central part). The meaning of these notations may be elucidated in the following way. Let us put three pancakes on top of each other in a saucer and then cut the pancakes

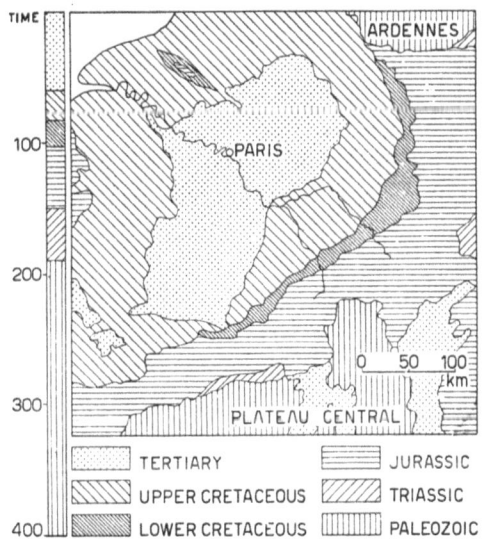

Fig. 5. Simplified map of the basin of Paris; the accompanying time-scale shows the approximate time of origin of the strata

off at the level of the upper margin of the saucer. If the bottom-most pancake has a blue colour, the middle one green and the upper one

yellow the resulting configuration is roughly comparable with the structure of the basin of Paris as expressed on the geological map of France
(fig. 5). Again the different colours represent different periods of time. The
yellow colour (dots) in the centre means that the rocks originated at the
same time as those which underly the Mythen. The surrounding green
ring dates from an earlier period when the limestones south of Brunnen
originated. The time of origin of the rocks indicated by the blue ring
(= horizontal shading) corresponds with part of the Mythen, which are
in part still older. The colours represent internationally accepted indications of special geological periods or time-intervals.

However, we need not travel to France if we want to study a basin-
shaped structure. Here, in Haarlem, we are in the centre of such a depression in the earth's crust: the so-called North Sea basin. Its surface layers
are at present partly above sea-level, partly below it. The foundation
below has gradually subsided and at the same time the saucer-shaped
depression became filled up with sediments, mainly erosion products
from the surrounding areas. The average rate of subsidence and sedimentation was less than ¹/₂ cm per century. However, the downward movement
started about 220 million years ago, which means that the bottom of the
basin is at present some 9000 metres below us. The North Sea basin is
separated from the basin of Paris by a vault-shaped ridge of very old
rocks, the so-called massif of Brabant, which occupies a great part of
Belgium. The eastern boundary of the North Sea basin is formed by a
similar though much narrower ridge-shaped structure in the basement
which is indicated by the name "axis of Erkelenz". Remarkably enough
it coincides approximately with the political eastern boundary of the
Netherlands (fig. 35).

I mentioned already the frequent occurrence of basin-shaped depressions. As a matter of fact, the earth's surface is as it were pock-marked by
a great many such basins. Fig. 6 gives an illustration, which expresses
yet another remarkable feature of the basins. The different notations
indicate different times of origin of the basins. The time of origin, i.e.
the beginning of the downward movement appears to be correlated with
an epoch which elsewhere manifests itself by widely spread mountain
making. Consequently, periods of origin or rejuvenation of many basin
shaped depressions are at the same time epochs of very intense mountain
making. Basins originating simultaneously with the Alpine chains are
marked by dots in fig. 6. Basins which originated at the same time as the
Ardennes and Appalachian chains have a black notation, etc. If we note
them on the time-scale the parallelism appears clearly. Again the rhythm

of 250 million years is found. Epochs of less intensive mountain making have been fixed at intervals occurring within the major rhythms. These, too, have their counterpart in the formation of basins. An example is shown by the basins with the diagonal notation in fig. 6. The corresponding epoch is marked by one of the smaller peaks of the graph in the corresponding column of fig. 8, viz. by the peak at 300 million years ago.

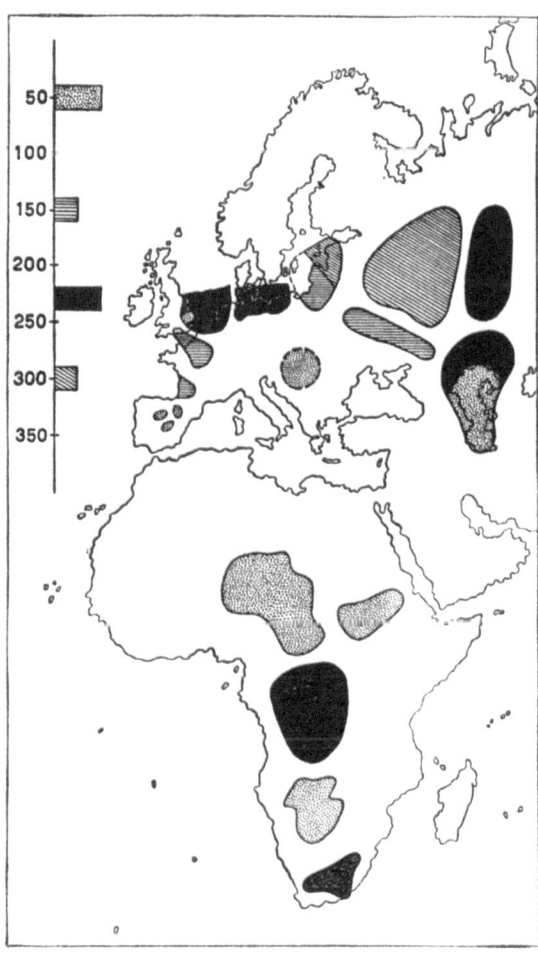

Fig. 6. The different notations in the European and African basins shown on this map indicate different epochs of origin as indicated on the accompanying time-scale.

Several basins originated at great distances from contemporaneous mountain-belts. Obviously, the processes causing the origin of mountain-chains were not restricted to the comparatively narrow belts but they had a world-wide influence on the earth's crust. It is a difficult problem, however, to ascertain which were the forces and processes that caused these events during certain specified periods.

Yet another problem is to explain why a mountain-chain rose up in a certain area whereas a saucer-shaped depression started its movement in another region.

It is not surprising that the still mysterious processes which affect the earth's crust, also find their expression in a rhythmic cadence of the sea-level. Repeatedly the sea has invaded large portions of the continents and then subsequently retreated. This chapter of the earth's history is an extremely monotonous one, but this very monotony, this continuous recurrence of the same events, constitutes one of its most important aspects. A graphic review of these events clearly shows their rhythmic occurrence and the time-relations with the events expressed in the two previous curves.

Apart from influences acting from deeper realms the sea-level is affected by other factors. It can easily be calculated that the level of the sea would be raised some 40 metres if the ice incorporated in our present glaciers and ice-caps over the whole world were to melt.

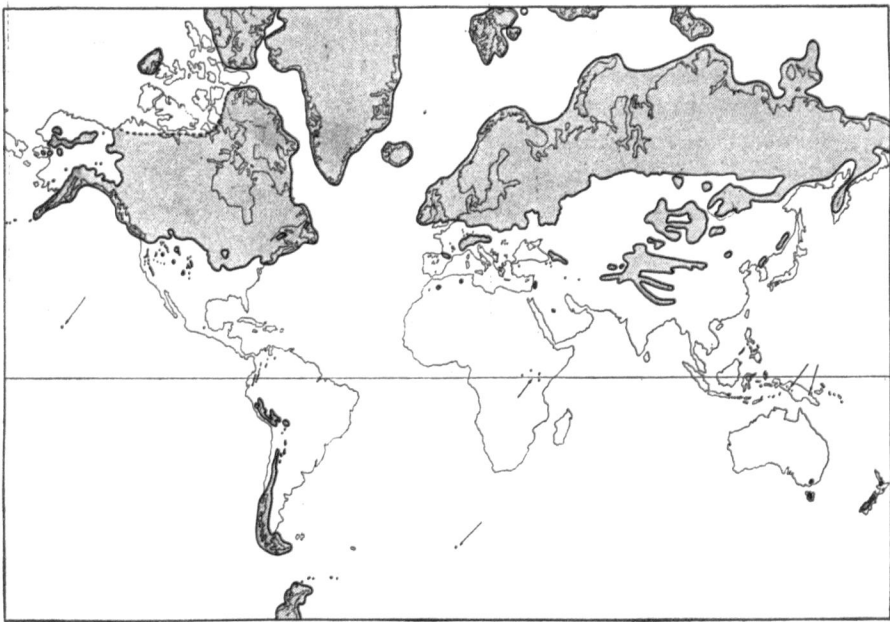

Fig. 7. Maximum extension of ice-sheets on the continents during the last glaciation.

Our present glaciers, as well as the ice-sheets on Greenland and Antarctica are, however, poor remnants of a once much bigger and wide-spread ice cover. Less than one tick back on the geological clock large parts of northern Europe, Asia, Canada and North America were covered by ice-sheets of enormous dimensions. An extensive elongated ice-mass rested on the southern part of South America, another on New Zealand, etc. (fig. 7). The water incorporated in these ice-masses was ultimately extracted from the oceans. As a result sea-level was about 100 metres lower than at present.

Ice-caps appear and disappear. Ice-ages have occurred only occasionally during the long history of our planet. During the long time intervals between two periods of glaciation the earth was free of ice and a more uniform and milder climate extended up to high latitudes.

We are living in an abnormal climatological age which might be called an intervening stage of milder glaciation or partial deglaciation. The actual distribution of pronounced climatic zones as well as the distribution of fauna and flora in distinct biogeographical units is geologically speaking a phenomenon of very recent origin. The last period of glaciation began about one million years ago. If we reach back some 250 million years we find evidence of the previous extensive glaciation. The same phenomenon is again found in even remoter times if once more we reach back for the same time interval. In short, if noted in a proper order on the geological time-scale, periods of glaciation and periods that were characterized by a mild climate up to a short distance from the poles, will show a large climatic periodicity (fig. 8). The time relations with the major rhythm of mountain building and other rhythmically occurring phenomena clearly show that in some way they must be genetically related. The terrestrial phenomena are interwoven as in an enchaining fugue. But there is still more of entrancing interest.

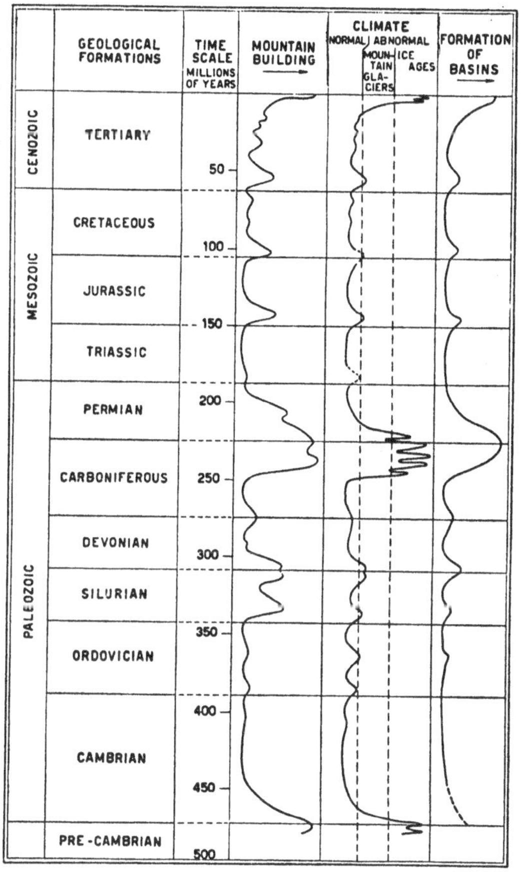

| GEOLOGICAL FORMATIONS | | TIME SCALE MILLIONS OF YEARS | MOUNTAIN BUILDING | CLIMATE NORMAL/ABNORMAL | | FORMATION OF BASINS |
|---|---|---|---|---|---|---|
| CENOZOIC | TERTIARY | 50 | | | | |
| MESOZOIC | CRETACEOUS | 100 | | | | |
| | JURASSIC | 150 | | | | |
| | TRIASSIC | | | | | |
| PALEOZOIC | PERMIAN | 200 | | | | |
| | CARBONIFEROUS | 250 | | | | |
| | DEVONIAN | 300 | | | | |
| | SILURIAN | | | | | |
| | ORDOVICIAN | 350 | | | | |
| | CAMBRIAN | 400 | | | | |
| | | 450 | | | | |
| | PRE-CAMBRIAN | 500 | | | | |

Fig. 8. Graphic representation of corresponding periods of (1) mountain-building, (2) formation of basins, and (3) periods of glaciation and epochs of minor occurrences of ice.

Let us consider another group of phenomena viz. the volcanoes which are so fascinating even to the non-geologist. One of the first questions is why volcanoes originate in specified areas and in special periods. The answer may differ according to the different areas considered. In most

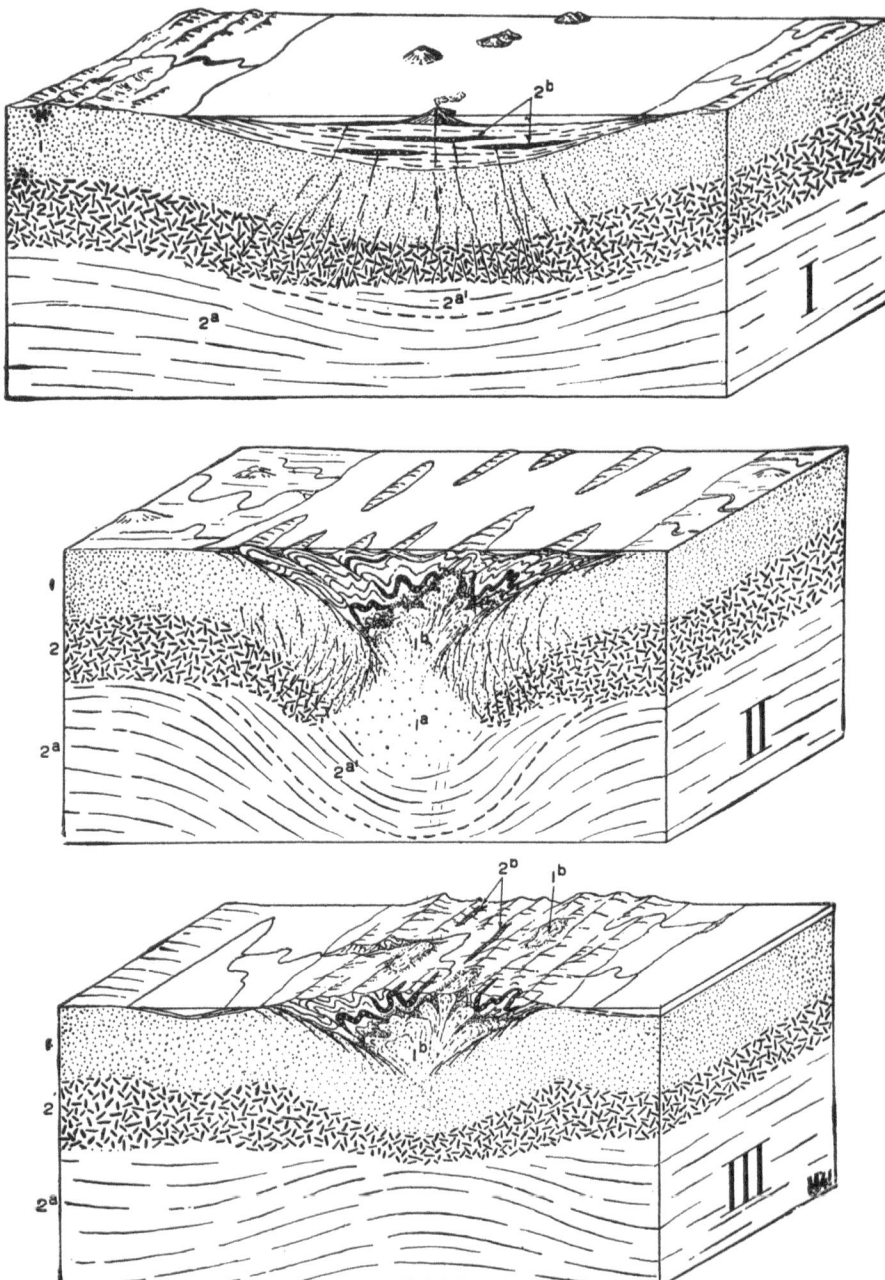

Fig. 9. The same block-diagram as represented in fig. 3 showing, however, also deeper realms of the earth's crust and its substratum as well as the accompanying volcanic and plutonic phenomena. Details are discussed in chapters V and VI. 1 and 2 crystalline crust; 1 granitic layer; 1a disintegrated part of 1; 1b acid batholithic intrusions and metamorphic aureole; 2 basic layer, 2a basic substratum; 2b basic abyssolithic injections, extrusions and volcanism.

cases, however, a relation to the rhythmic origin of folded mountain chains can be traced.

I mentioned already that mountain-chains are born from elongated belts which started as downward moving strips in which sediments accumulated in abnormal thickness. When the downward movement reaches a certain limit, the strength of the earth's crust will be exceeded and disruption will follow. Earth material from below the crust will then immediately fill the fissures and faults in the crust. Appearing at the surface it will form outflows or build up cones of ejected material, which we call volcanoes (fig. 9, block 1).

In a later stage the subsiding strip of the crust will bend so far that it collapses. One of the consequences will be that it will buckle downward so as to form a root of crustal material penetrating into deeper realms of progressively higher temperature and pressure. Disintegration of the root and partial upward migration is shown tentatively in block II of fig. 9. This material gradually forces its way towards the surface, eventually penetrating into the folded contents of the sedimentation trough, including the earlier erupted and intruded masses. As a matter of fact, the phenomena involved are very complicated. What matters for the moment, however, is only that these processes are quite different from those which are associated with material coming up along fissures and faults during an earlier stage of development of the subsiding trough. The bulk is much greater, but the movements are very much slower and come to a standstill before the surface is reached. They are called plutonic processes (fig. 9, block II).

Some areas show yet another relation between volcanism and folded mountain-chains. Parallel with the zone of buckling two belts of deformation may develop in the earth's crust. The upward undulation accompanying the zone of buckling on its concave side is, as it were, predestined to become a volcanic arc. For the arching of the vault induces relief of pressure at the lower side of the crust. Accordingly, earth material from below the crust ascends in these parts along deep-reaching cracks and gives rise, by intricate processes, to volcanic phenomena. At the same time faults and rift-valleys develop in the crest of the vault and facilitate the building up of volcanic bodies on the surface. Fig. 10, a schematic and tentative block-diagram of the East Indies, shows these relations between the zone of buckling and the row of volcanoes which can be followed over Sumatra, Java and the Lesser Sunda islands. To the left of the volcanic belt, the next downward warp of the crust is represented by a series of troughs which have been partly filled up with sediments and were folded subsequently. Two series of elongated troughs accompany the outer arc (which coincides with the zone of buckling) on either side. Their origin is probably due to another process. When, at a later stage, compression in the crust decreases the comparatively light rocks that form the root in the zone of buckling will show a strong tendency to rise. In this process the zone of the mountain

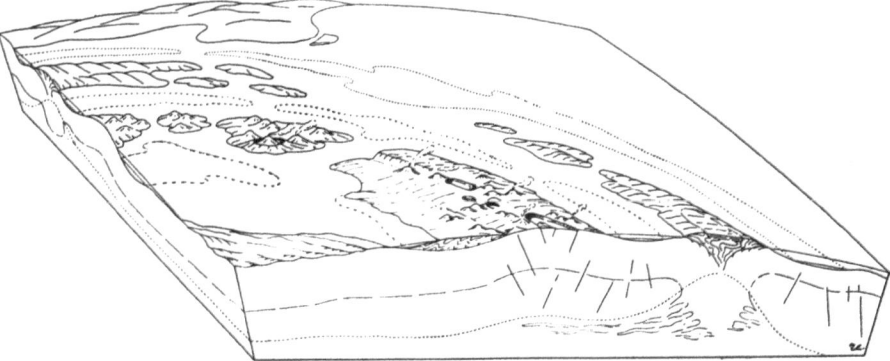

Fig. 10. Schematic and tentative block-diagram of the East Indies, showing the root of light material below the outer arc, the inner arc with its volcanoes, and the different zones of troughs.

root will probably detach itself more or less from the adjoining areas. At the same time these areas, which were first elevated by the buoyance of the root, must sink back. Moreover, material below the crust will flow towards the rising root and, as a further consequence, a furrow will form on the earth's surface on either side of the buckled belt.

But let us return to the volcanoes. Volcanoes may originate in very different areas. I shall mention only one further possible mode of origin. Basin-shaped depressions in the earth's crust, which were discussed a few minutes ago, have their counterpart in dome-shaped elevations. An example is the so-called Rhine-shield (fig. 35). Deep-reaching cracks (so-called *faults*) affected the vault as a result of its updoming movement. Often rift-valleys or graben originated i.e. comparatively low lying blocks separated from the adjacent parts by steep faults. In this way the Rhine rift-valley separates the Vosges and the Black Forest. Other examples are the Plateau Central with the Limagne-graben, and the African-Arabian shield with the rift-zone of East Africa and the Red Sea. Along the faults volcanism was offered an opportunity to manifest itself (fig. 11). Probably many among you have seen one or more of these volcanic areas: the hardly extinct volcanoes of the Eifel and Rhön, or the equally well preserved volcanic cones and lava flows in the neighbourhood of Clermont-Ferrand: Pariou, Puys de Dome, etc.

One of the most remarkable features of the dome-shaped elevations is that the updoming movement is related in point of time to epochs of mountain-building and the origin of basin-shaped depressions (fig. 18). Again the same major rhythm is found, evidently pointing to a genetic relation between all these phenomena which at first sight are so heterogeneous.

During periods of decreasing compression in the earth's crust, a folded zone may rise so as to form a mountain-chain; elsewhere a basin-shaped depression forms; in another area a dome-shaped elevation originates.

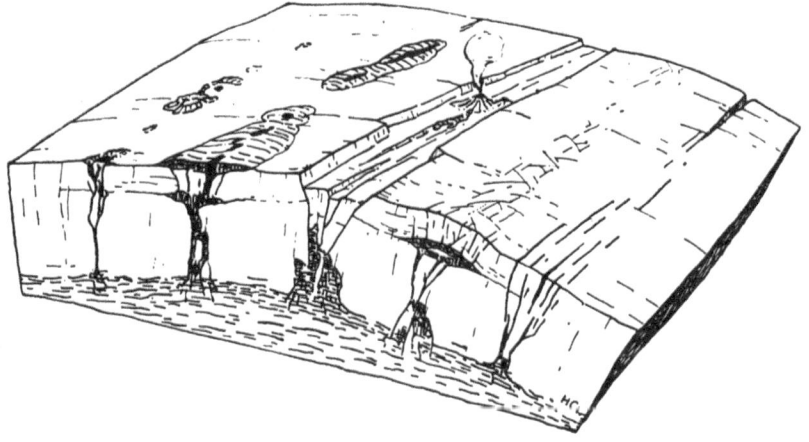

Fig. 11. Schematic block-diagram of a dome-shaped elevation and accompanying volcanism (After H Cloos).

The problem is why these different structures and partly opposed movements occur in the specified areas where they actually are found. The processes are probably controlled by two major factors. One is the influence of primeval structural elements of the earth's crust. The other concerns the displacement of material below the crust. However that may be, you observe how our terrestrial symphony has become enriched by a sixth motive, which is harmoniously interwoven with the other ones.

Along the six different routes of volcanism, sea-level, glaciations, folded mountain-chains, basins and dome-shaped elevations, we arrive at problems of the depths of the earth. What is called the earth's crust is a thin cover of crystalline rocks (fig. 12) — their thickness being not more than one hundredth part of the earth's radius — which is supposed to rest on non-crystalline and denser material, the so-called substratum.

Approximately 2900 kilometres below us is the outer boundary of the terrestrial nucleus, which was generally thought to consist of nickel-iron,

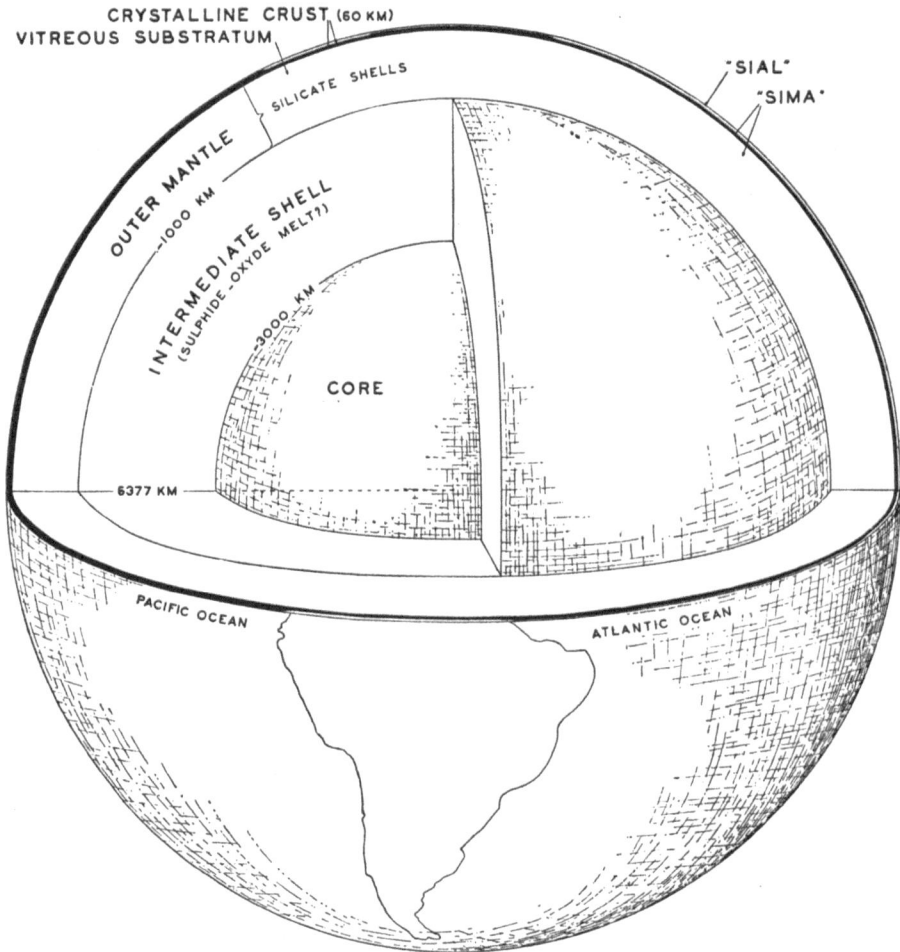

Fig. 12. Hypothetical picture of the earth's interior

but which is now considered by some geophysicists as a mass of still undifferentiated solar material.

And what is the age of the earth? Similar results as regards the age of the earth were obtained by investigators in widely different branches of science and show that some 3000 million years have elapsed since its origin; 3350 million years is the latest figure calculated. Moreover, this period seems to have formed not only the beginning of earth history, but it appears to have been a critical date in the whole universe which since then has developed gradually into its present condition. Apart from the earth, the moon, and the planets, the entire solar system received the impulse to perform its present movements at the same moment. Even more than that. The solar system is but one out of ten thousand to a hundred thousand billion stars which together form our galactic system or spiral galaxy. Spiral galaxies are spread fairly evenly through space.

Fig. 13. The Andromeda nebula, a spiral galaxy

Some 40 million spiral galaxies appear to exist within the reach of the strongest telescopes. Supposing that — à la Jules Verne — we were to let ourselves be fired from the earth in a rocket travelling at a rate of 5,000 miles an hour, it would take us no less than five thousand million years to pass through a spiral galaxy such as the so-called Andromeda nebula (fig. 13).

The most distant galaxies — representing the very extremities of that part of the universe which is known to us at present — are separated from us by a thousand million light-years. This means that if it were at present possible to scan the earth through a super-telescope from one of these extremely distant parts of the universe we should find that we were looking at the first and most primitive organisms in our terrestrial history, swimming around in primeval seas, and we should have to wait 700 million years to detect the first signs of life on the continents — that is, assuming that the distance remained unchanged during this lapse of time. However, one of the most sensational astronomic discoveries of the twentieth century is that all the galaxies appear to recede from our own galactic system, as well as from one another. By reversing the picture and imagining the galaxies to travel towards instead of away from one another, we are able to conclude that some 3000 million years ago a tremendous quantity of matter was packed into a small volume of space. Evidently, the chances for the origin of galaxies, solar systems, comets, and planets must have been great at that time.

Our galactic system rotates around its centre. Now the remarkable fact is that it achieves a full rotation within a time-interval which you know already from so many terrestrial processes and which now may even sound familiar to you: about 250 million years (fig. 18).

Perhaps the last named coincidence is a question of pure chance. As a matter of fact it is impossible, at present, to demonstrate a cause to effect relation between the rotation of the galaxy and the rhythmic recurrence of processes in the earth's crust and its substratum.

At any rate the terrestrial symphony has risen to celestial regions. It leads to speculations on the aeons of time, on the incomprehensible dimensions of curved space, and on the meaning of life and matter which it embraces. The fugue soars to such a giddy height that we ask where it will end.

Did you ever notice — when listening to a symphony — how a composer manipulates his theme at such a moment? Suddenly he returns to his first motif.

*Tema come prima e Finale.*

Therefore, let us once more look at the Mythen rocks.

When showing you the first picture, I was very positive in asserting that the rocks of the Mythen are much older than the mountains on which they rest. Still, a geologist with his hammer is not like a magician and his staff. The rock fragment which he knocks off with his hammer and which he carefully examines — first in the field with his pocket-lens, afterwards at home under the microscope — may contain remains of organisms that lived in a long distant past. Taking into comparison present day conditions, he tries to make out whether the rock was formed on land or in the sea, in a desert or in the tropics, in shallow water or in the depth of the ocean. But there is yet another puzzle which he tries to solve. Gradually, data assembled from all parts of the world, have revealed an ever stronger

Plate I

(A) Photograph of a fossil Ichthyosaurus from the Lower Liassic of Holzmaden (specimen in the Institute of Mines at Delft).

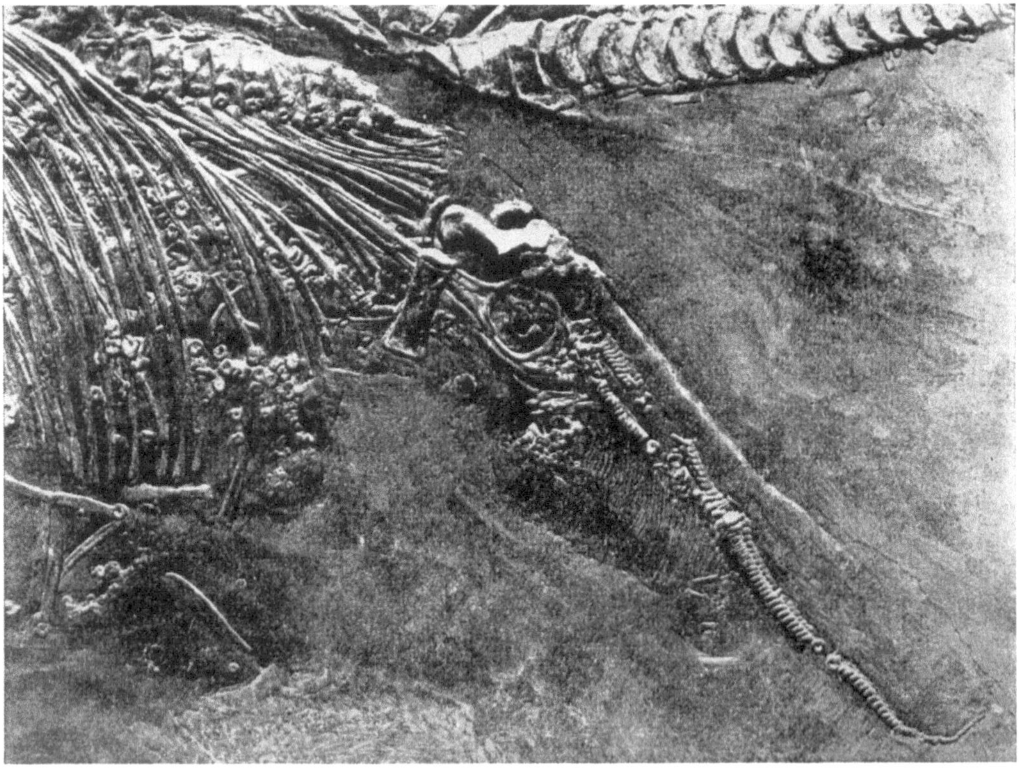

(B) Birth of an Ichthyosaurus, its snout still fastened in the pelvic bone of its mother. (After B. Hauff).

evidence of the evolution of plants and animals during the long duration of the earth's history. Biology and comparative anatomy are of undispensible value in studying fossil remains of organisms. The results attained so far are of such comprehensiveness that I cannot possibly do more than mention a few examples.

When the limestones of the Mythen were deposited the supremacy among vertebrate animals was with the reptiles, on land as well as in the air, in rivers as well as in the seven oceans. Sometimes their remains have been found so perfectly and completely preserved that hardly any reconstruction is needed. Plate I (A) portrays an Ichthyosaurus which, so to speak, swims past you like a fish in an aquarium. All parts of the skeleton are in their right place. The skin is preserved. Dorsal fin and tail are exactly *in situ*. We know what food they liked and how the injuries of their bones healed. They were viviparous and very probably the snout of the young animal, when being born, remained for some time firmly fastened in the pelvic bone of its mother. In this way the young Ichthyosaurus was dragged along on its mother's journeys until it had grown strong and fit enough to swim full speed on its own (Plate I, B). Among present day mammals dolphins appear to have the same habit.

Details like these can only be known in exceptional cases when specimens are found in the position shown by Plate I (B). Generally, however, finds are not so self-evident. The haphazard accumulation of *Stegosaurus* bones, shown in fig. 14 (A), have been carefully mapped before they were excavated and mounted. Additional data were derived from other finds of similar creatures. Moulds of the skin were found occasionally. The set of teeth was analysed. A detailed and comparative study was made of the entire skeleton in order to reconstruct the correct position of the animal as well as of the mechanical possibilities of its movements. All remains of other animals and plants in the same strata were taken into account in order to reconstruct the original landscape. In this way hardly any imagination was needed to picture a Stegosaurus couple strolling around in search of lush leaves and mellow sprouts of young araucarias (fig. 14, B).

We even know some remarkable features of the nervous system of these creatures. Stegosaurus had very little brains. In proportion to dimensions and body weight it had to be content with 1/200 part of the weight of the human brain. Now, if one realizes how slowly human brains work — I mean in general, of course — and what queer sorts of thoughts come up in our minds — sometimes — one wonders about what a Stegosaurus may ever have pondered. Moreover, you will notice a thickening of its spinal column in the pelvic region (fig. 14, C), a thickening which surpassed

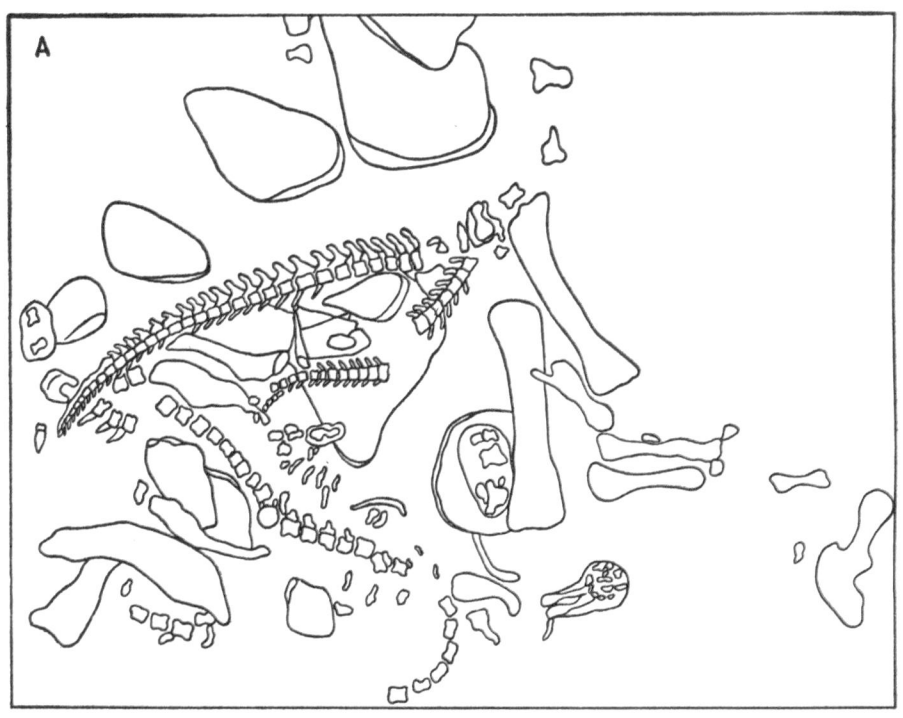

the size of the brains. So, if ever this creature had a luminous idea, it more probably came up in its hind quarters than in its head.

The largest beast of the world used to be on show in Berlin: I mean the complete skeleton of a Brachiosaurus (fig. 15, A) which was also an animal with very poor brains. A skilful analysis of the skeleton, which must have been more absorbing than a detective story, led to the conviction that Brachiosaurus and his relatives like Brontosaurus (fig. 15, B) had a preference for dwelling in large rivers and swampy regions, more or less like hippopotami in our days.

Naturally, finds of enormous creatures like Dinosaurs do not belong to the daily adventures of a geologist. Much more often he has to occupy himself with remnants of small or even very tiny organisms. This does not alter the fact that their structure is often very complicated and a reconstruction demands much time as well as patience and deliberation. Sometimes the skeleton of the animal can be studied only under the microscope in a number of arbitrary thin-sections cut from the rock in which it happens to occur. The model of the creature, represented in fig. 16 and having a diameter of four millimetres, has been reconstructed by such a procedure. A few years afterwards complete specimens were

Fig. 14. (A) Haphazard accumulation of Stegosaurus bones found in the Morrison beds near Como, Wyoming (After Ch. W. Gilmore).
(B) Reconstruction of a Stegosaurus couple in their original surroundings (After Charles Knight).
(C) The nervous system of Stegosaurus (After H. G. Wells, J. Huxley and G. P. Wells).

Fig. 15. (A) Brachiosaurus skeleton in the geological Museum of Berlin. Its enormous dimensions can be appreciated by comparison with the human skeleton to the left (Drawing after a photograph).

Fig. 15. (B) Reconstruction of a Brontosaurus in its original surroundings (After H. G. Wells, J. Huxley and G. P. Wells).

found elsewhere in a marl from which they could easily be picked out and studied in every position required. The reconstruction was checked and proved to be correct.

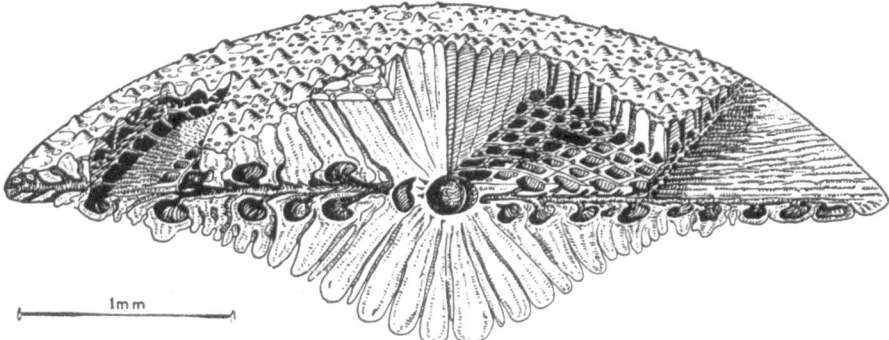

Fig. 16. Reconstruction of the skeleton of Biplanispira, an extinct marine animal having a diameter of 4 millimetres.

Such small organisms are often used in determining the relative age of a rock [1]). This means that we must make out whether its organic remains show it to be either of the same age or older or younger when compared with another rock of known age. A decision of this kind is possible only because plants and animals passed through an evolution. It is because of their evolution that fossil plants and animals furnish a means of determining the relative age of a rock, i.e. the time of its deposition or sedimentation. This principle is of paramount importance to geology. This statement needs a short historical explanation.

In the time of Newton the general opinion was still that fossils were images made in stone by a mysterious formative power, called *vis plastica* or sometimes *spiritus architectonicus*. One of the last exponents of this wholly mediaeval doctrine was the unfortunate Beringer, who was such a convinced supporter that it was only after the publication of his *Lithographia wirceburgensis* (1726), in which he gave illustrations of such remarkable "fossils" as comets, spiders in their webs, letters, etc., that one day he discovered his own name in stone, and only then realized that he had been the victim of a student's practical joke.

After that came the period of the so-called "Diluvianists", who did

---

[1]) The absolute ages indicated in the graphs, fig. 4, 8, and 18 are based on the fact that occasionally igneous rocks contain rare radioactive minerals which, due to the breaking down of elements like uranium and thorium into stable endproducts at a known rate of production, allow a determination of their absolute ages in millions of years.

regard fossils as remnants of organisms, but thought them all to have come from one great inundation, the biblical deluge. Famous is the find of the Swiss Scheuchzer, who thought that he had found the fossil skeleton of an old sinner who was drowned in the deluge, a *Homo diluvii testis* in a slab of limestone from the quarries at Oeningen (1711). Some cast doubt on his opinion and considered the find to be not the skeletal remains of a human being, but those of a large frog or other amphibian. Ultimately, when Cuvier in 1811 visited Teyler's Institute in Haarlem, where the famous piece was preserved (and still is), he laid bare the whole skeleton from the stone; it turned out indeed to belong to a very large amphibian, some such animal as the Japanese giant salamander (Cryptobranchus japonicus).

The investigations of two scientists, Cuvier and Smith, who both happened to have been born in 1769, put an end to this "school" too. When William Smith in 1831 revealed how the successive strata in England are characterized by a definite sequence of organisms, *stratigraphy* was born, a branch of science which concentrates on the study and description of sequences of strata all over the world and the fossils they contain. It forms the foundation of comparative regional and historical geology. Since then earth science, which is now barely a century old, grew at such a pace that the classic strife between the so-called Neptunists and Plutonists seems to us as far back in the past as the Wars of the Roses, although it is less than a hundred years ago that the followers of Werner and Hutton fiercely attacked one another in their writings.

The insight which William Smith had given made it possible to study the history of organisms in conjunction with that of the earth, while Cuvier at the same time was the founder of the comparative anatomy of fossil animals. All this made science receptive for the following important step, the theory of evolution.

Fossils provide the only means of deciphering earth history in some detail. Organic remains are known in appreciable quantity only from strata which were formed during the last 500 million odd years. Consequently, reconstructing the course of events in still earlier times the deposits of which are devoid of clear fossils generally offers insurmountable difficulties. Therefore, the curves of fig. 4, 8 and 18 have been limited to the last 500 million years.

For the same reason geology, the youngest among sciences, started its course not earlier than about a century ago, when it had become clear that organic remains in rocks occur in a special regular sequence of faunas

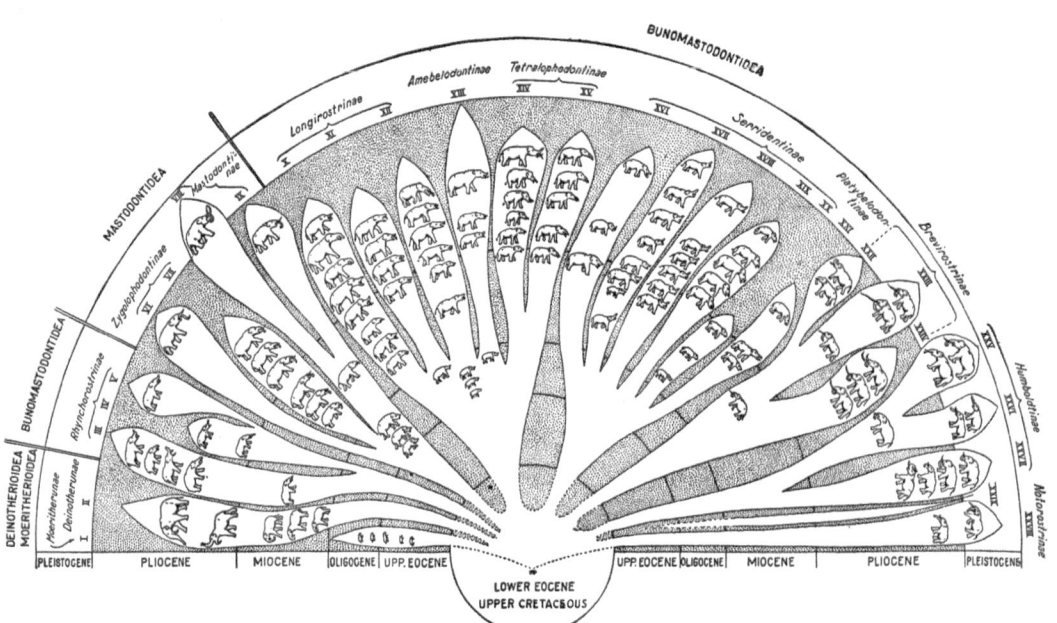

Fig. 17. Schematic pedigree of the Mastodonts (After H. F. Osborn).

and floras, and not as chaotic assemblages. On the other hand the correctness of the principle of evolution is proved daily by the practical results obtained from the determinations of the relative ages of sediments.

In the case of the Mythen and their underlying strata the age determination is partly based on organisms which were much smaller even than the marine animal of fig. 16. Moreover, the latter lived in the East Indies and the Pacific at approximately the same time when the "yellow" strata of the surroundings of Brunnen originated.

In the meantime these questions lead us to considerations on the distribution and relationship, and consequently on the descent of organisms. As an example of the latter subject fig. 17 shows a fan-shaped pedigree of the Mastodonts. This tentative pedigree comprises 1,200,000 generations, though — it must be granted — not all are known by name and surname. It extends over about 50 million years, i.e. the lower part coincides with the time when the "yellow" strata on which the Mythen are now resting began to accumulate in a shallow sea, and the whole pedigree extends over at least three times the span of time needed for the deposition of the "yellow" sediments.

The long history of a certain group of organisms, as revealed by studying their fossil remains, in many cases shows a remarkable analogy to phenomena which we know very well from living beings, including ourselves. Moreover it shows some aspects which we could not possibly learn from other sources. Let us give an example of both features. Consider the development from seed to flowering plant, from caterpillar to butterfly, from baby to full-grown man; consider in general the harmonious growth-process of a living being. It shows numerous phenomena interwoven with an inherent efficiency in a striking coordination of function and vital processes, and a sequence of events with a marked direction or linearity.

Now, a marked direction, for example, finds also its clear expression in the long enduring evolution of a group of organisms, which might be called their geological history. The evolution of the Mastodonts shows a more or less rectilinear trend or direction in several features: e.g. increase of body dimensions, perfection of molars, growth of tusks, jaws, and trunk. Apart from this phenomenon, however, the pedigree of the Mastodonts shows that several lineages, part of which evolved in different continents, present similar tendencies though at varying rates of evolution.

If one becomes absorbed in these problems one automatically arrives at the question: what is the fundamental difference between a living and a non-living system?

In search of an answer one route leads to the terrain of modern atomic physics, another route to biochemistry, the third route compels us to

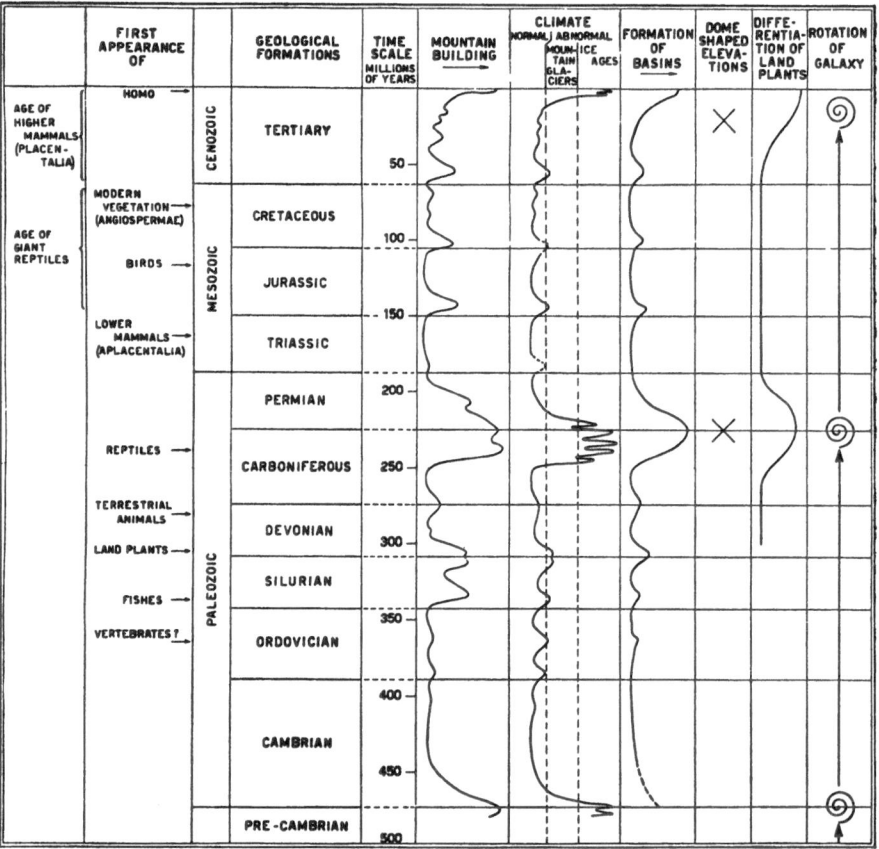

Fig. 18. Graphic representation of part of the symphony of the earth showing the time relations between (1) epochs of mountain-building, (2) glaciations, (3) formation of basins, (3) dome-shaped elevations, (5) epochs of differentiation of the flora and (6) rotation of the galaxy.

study the mechanism of inheritance, the fourth field of research belongs to microbiology. Last but not least the fifth route brings us to the history of the organic world during the long eras of geologic time.

Evidently, the latter is an indispensable source of information, inasmuch as it reveals a set of phenomena with which we cannot possibly become acquainted along    different path of exploration. One of the results obtained time after time is that the evolution of plants and animals was influenced by their organic and non-organic environment. Two examples may illustrate this latter influence.

During two periods of earth history land plants became differentiated into a number of different botanical provinces. These periods of differentiation of the flora coincide with the two periods of world-wide mountain-building and large spreading of continental ice-caps, which are separated from each other by the last time interval of 250 million years (fig. 18). Surely, the coincidence of the two peaks of plant evolution with the two peaks of environmental changes is not a question of pure chance, for coincidences of a similar kind are too remarkable and occur too frequently. To finish, let me therefore mention a second example.

With the last rising up of mountain-chains, when large ice-sheets spread over the continents, and when sea-level fell 100 metres below its present position, with this last tick of the geological clock — when more than 3000 million years had passed without witness by human beings and when the galaxy had revolved twelve times around its centre — there begins an era which is called the *Psychozoic*. This term received its name from the fact that Man, the specialist of spiritual and intellectual differentiation, appeared on the stage and began to extend his supremacy over the world.

Simple facts like these make us feel humble and unpretentious. Moreover they give us something to think over concerning the past as well as the future of mankind.

Thus, at last, we arrive at Homo sapiens and his problems. Homo sapiens: whose history reaches not farther back than the time when the Alps rose to their present altitude, the time when the landscape that formed our starting-point began to become moulded into the scenery which we admired in the surroundings of Brunnen and the Mythen. Everybody enjoys looking at this beautiful scenery from aesthetic motives. However, I hope to have made clear that for a geologist the same landscape may — moreover — arouse a number of far reaching speculations.

And thus we reach the end of the terrestrial symphony the score of which has been skimmed very briefly indeed. It started with a motif full of unrest. It ends in a final major chord, expressing the grateful feelings of a scientist who realizes his privileged position of being able to devote his life to studying the vast realm which extends from the remotest ages and depths of the universe to the origin and meaning of all organisms including the inmost depths of ourselves.

## REFERENCES

Chapter I is mainly translated from an address to the Dutch Society of Sciences (*Hollandsche Maatschappij van Wetenschappen*) at Haarlem, May 17th 1947, and published in the non-illustrated series *Haarlemse Voordrachten* no. VII (De Erven F. Bohn N.V., Haarlem 1948) under the title *Symphonie der Aarde*.

In order not to be deflected from the main outline that had to be followed no names of authors were mentioned and details were omitted altogether. Those who are interested in a more elaborate account will find some subjects treated more extensively in the following chapters and still more so in the following two books published by the present author at Martinus Nijhoff's, The Hague: (1) *The Pulse of the Earth*, (2) *Leven en Materie* (Life and Matter).

# CHAPTER II

# A COUNTRY BELOW SEA-LEVEL

## INTRODUCTION

*The problem.*

Everybody knows that Holland is a flat country. And one of the most characteristic features of this small corner of Europe is that its surface is a few metres below sea-level. The low lying western provinces of Holland are separated from the sea by a rather narrow strip of dunes, which came into being at about the middle of the ninth century. Since those times the sea developed into an ever more threatening factor and the Dutch people became involved in a constant battle with the sea which is considered its arch-enemy. Dykes and stone groynes had to be built along the coast and the land immediately behind the dunes had to be drained; canals and other water-works, in short a whole system of so-called *polders*, came into being.

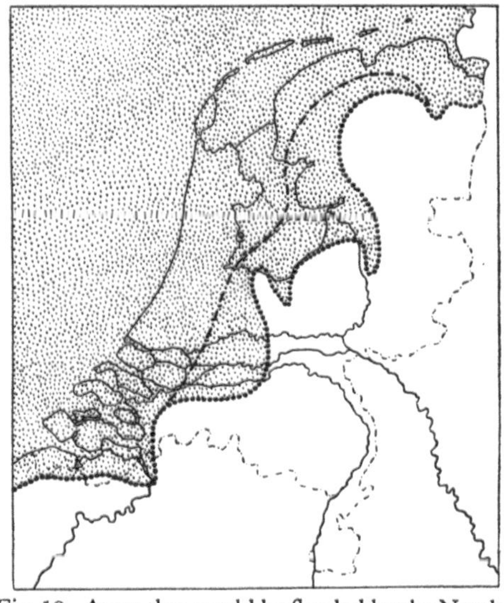

Without these artificial constructions great stretches would be flooded by the sea (fig. 19).

I want to make clear what is the cause of the 1000-years battle of the Dutch people with the waves of the North Sea and why the battle has become especially acute since the ninth century.

Fig. 19. Areas that would be flooded by the North Sea if not prevented by artificial constructions. The dot-dash line indicates the extension of old sea-clay.

This means that I am going to tell you the story of a complex of sand, clay and, last but not least of a thin layer of peat.

At first sight such a story will appear to you very uninteresting. As a matter of fact, however, it is a most fascinating problem, a sort of detective puzzle. A first complication presents itself in that actually there are two sand masses and two clay layers of different age separated by an intervening peat deposit. Their position in the coastal district of Holland is shown diagrammatically in fig. 20. Fair sand dunes along the coast rest on a series of parallel sand barriers which, of course, were formed at

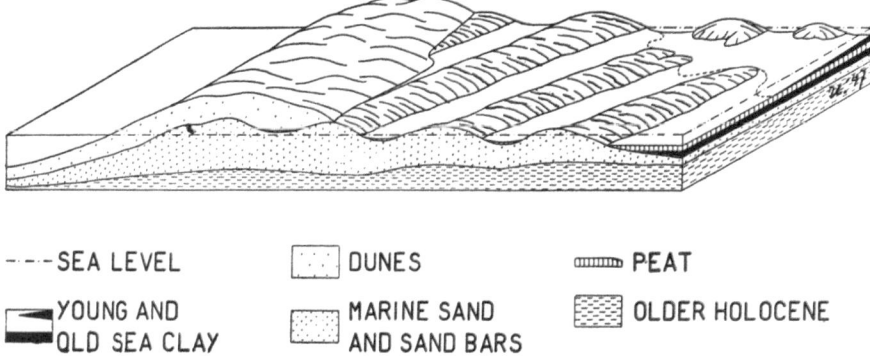

--- SEA LEVEL          DUNES          PEAT

YOUNG AND          MARINE SAND          OLDER HOLOCENE
OLD SEA CLAY          AND SAND BARS

Fig. 20. Schematic block-diagra ᴇ showing the geological formations in the coastal district of Holland.

an earlier time than the dunes. The clay layers are marine deposits, called old and young sea-clay respectively. The dunes and the young sea-clay were not yet present in Roman times. Both originated after the middle of the ninth century. The other elements were already in existence by then. The surface of this complex of clay and peat is now about two metres below sea-level. So, it seems as if the land moved downwards. Or was it the sea-level that did not remain constant? This forms one of the main points of our puzzle.

*The spit hypothesis.*

In a detective story it often happens that the reader first becomes acqainted with the official opinion of Scotland Yard. When the mystery seems disentangled and the solution seems quite clear, the private detective arrives on the scene and along a different route hunts down the real culprit.

In the same way I shall start by mentioning the once prevalent opinion about the origin of the Dutch coast. It is an opinion which sounds very obvious, viz. the coast is thought to have formed as a large sand bar which grew in a northerly direction from the moment the land connection between England and France was severed by the sea. There is a surplus of water movement along the Dutch coast in a northerly direction. Hence, the sand material of the dunes may have been derived from the demolition of rocks, which once filled the gap of the English Channel.

Marine clays — the so-called old sea-clay — accumulated in the bay behind the spit (fig. 21, A). Afterwards the bay became a freshwater lake in which a few metres of peat accumulated (fig. 21, B, C).

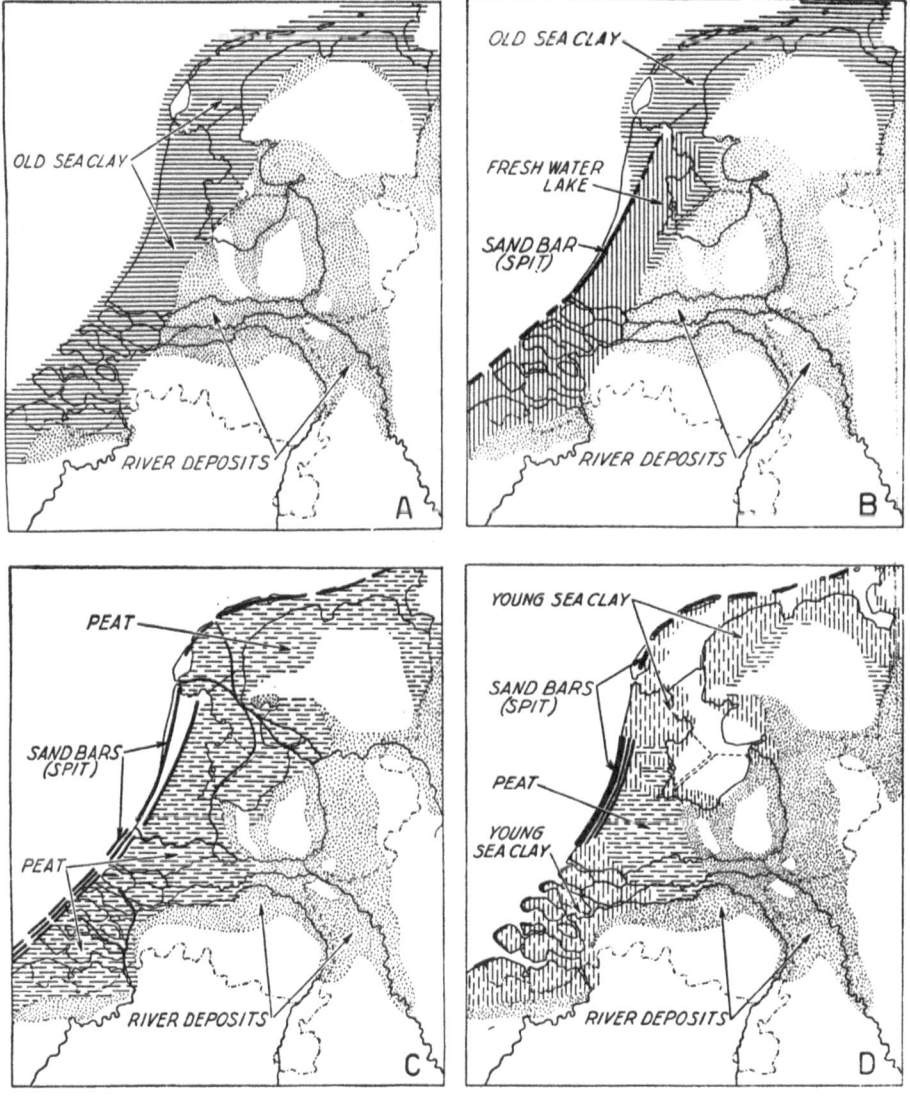

Fig. 21. Origin of Holland according to the spit-theory (After P. Tesch).

In a still later stage the sea broke through the sand bars. Its deposits are called young sea-clay (fig. 21, D). One of these marine invasions is known as the former *Zuiderzee*, which was artificially transformed into a freshwater basin, some fifteen years ago. It is now called *Ysel lake*.

A synopsis of this point of view may be seen in the graphic representation, fig. 23. It shows the thickness of the respective deposits, the supposed time of their origin and the situation of sea-level during the last 7,000 years. The supposed age of the formation of the two coastal elements is also indicated in the graph. This is the theory of "Scotland Yard".

*Unexplained features.*

A remarkable feature is that the parallel sand barriers are intersected abruptly by the coast south of The Hague (fig. 22). The theory which I showed you just now has no explanation of this phenomenon.

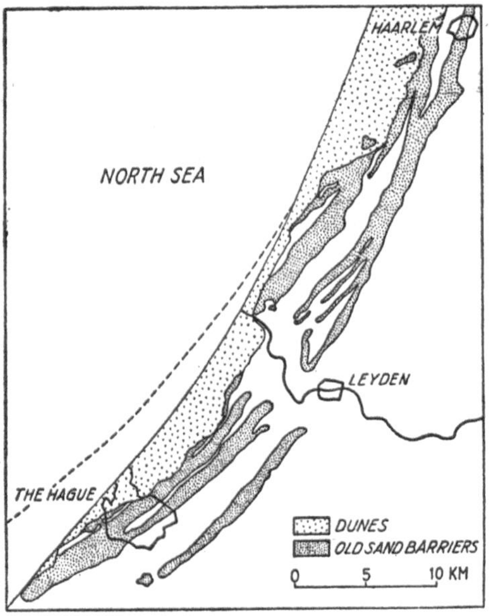

But there are several other serious objections against the theory which seemed so obvious at first sight.

Originally it was supposed that the material of the dunes was derived from the rocks that once formed the connection between France and England. Of course, the time of origin of the Dutch coast was thought to coincide with the time of origin of the open sea-connection between the Atlantic and the North-Sea; I mean the opening of Dover Straits. From that moment on the sand bar would have started growing. However, Dover Straits dates from an earlier epoch than the time of origin of the Dutch coast. Then, it was demonstrated in a convincing manner that no sand transport of any significance takes place from Dover Straits towards the Dutch coast at present. Therefore, it remains a puzzle whence the sand came that formed the dunes and what were the factors that caused the sudden accumulation of these dunes in the ninth century A. D. No satisfactory explanation was given by the theory of the sand bar.

Fig. 22. Abrupt intersection of the old sand barriers by the coast south of The Hague.

Moreover, investigation of bottom samples from the North Sea yielded a quite different result. For, it appeared that a great part of the dune-sand was supplied by land areas in the southeast, certainly not from Dover Straits and the English Channel. I shall return to this question afterwards.

But there is still another puzzle. According to the old theory the whole sequence of deposits accumulated during an uninterrupted rise of sea-level (fig. 23). However, several phenomena do not fit in with this con-

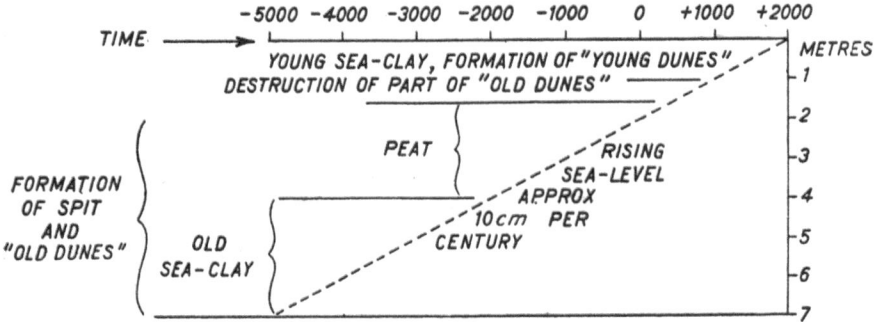

Fig. 23. Graphic representation of the origin of the deposits in the coastal district according to the theory of an uninterrupted rise of sea-level.

ception. How can we explain the change into freshwater conditions under constantly rising sea-level?

The situation is still more remarkable. The peat of the western provinces is no low level bog-peat that accumulated below a body of standing freshwater. Instead it belongs to the type called upland moor or *Sphagnum* peat, which means that it grew on a land surface. It is impossible to reconcile this point of view with the theory of a continually rising sea-level.

ORIGIN OF THE DUTCH COAST

*Material deposited by rivers and ice.*

On account of the difficulties and contradictions which are inherent in the spit-hypothesis, let us make a clean sweep and start all over again.

We will turn our attention to other parts of the Netherlands, without thinking about sand bars and dunes for a while. When tracing the surface-history of the country the coastal district will come up for discussion again, but then we may have a fresh and different view on its problems.

The Netherlands form part of a basin-shaped depression of the earth's crust. The subsiding movement dates at least from some 220 million years ago but it has been also an important factor in the most recent history. Large quantities of waste products were transported to this area by the rivers Rhine and Maas. And as the rivers wandered back and forth over considerable distances their deposits mingled over the greater part of the country (fig. 24).

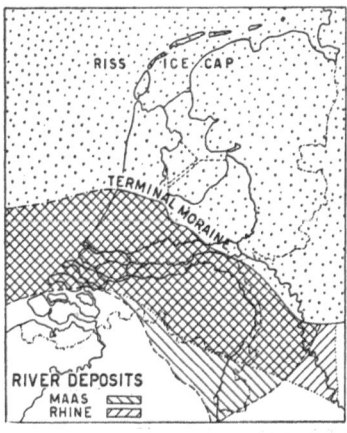

In the northern half of the Netherlands these deposits were covered during the climax of the Ice Age by the southern extremity of the Scandinavian land-ice cap, some 150,000 years ago. The terminal moraine of this land-ice cap is shown by fig. 24.

Glacial boulder clay extends over large parts of the floor of the former *Zuiderzee*. When the great dyke was built one of the most difficult problems was to tighten the last gap, which was held open by strong tidal currents. Without great quantities of boulder clay at hand the enterprise would have been much more

Fig. 24. The pre-glacial delta formed by waste products transported by the rivers Rhine and Maas and the terminal push-moraine of the land-ice cap.

troublesome and expensive. Now great lumps of resistant loam could be dredged from the sea-floor and dropped in the gaps. Contrary to sand this material was not loosened by the water and the heavy boulders remained in place, gradually to settle into a compact mass.

The general radial arrangement of ridges in the north eastern provinces of Drenthe and Groningen reflects the general trend of movement of the ice. And drumlins and glacial lakes are to be seen in Friesland.

But I must return once more to the frontal moraine (fig. 24). It is not a frontal moraine in the same sense as one speaks of the frontal moraine of a glacier in the Alps or the Himalayas. We call it a push-moraine. And, as a matter of fact, the land-ice has pushed together the pre-glacial river deposits at its front. This explains its comparatively great dimensions and height. In cross-section folding and thrusting of the pre-glacial layers can often be noticed.

The land-ice extended over the greater part of the bottom of the present North Sea and fused with another ice-cap which had its centre of origin in England and covered the British Islands with the exception of their southernmost part. Of course the water of the Rhine, the Maas, and melt-water of the ice-cap accumulated in a large lake in front of the land-ice (fig. 25). The water of the Glacial Lake had to find an outlet in a southerly direction. The erosive action of the water probably formed a channel which was widened and scoured out by the sea in later times to become what is now known as the Straits of Dover.

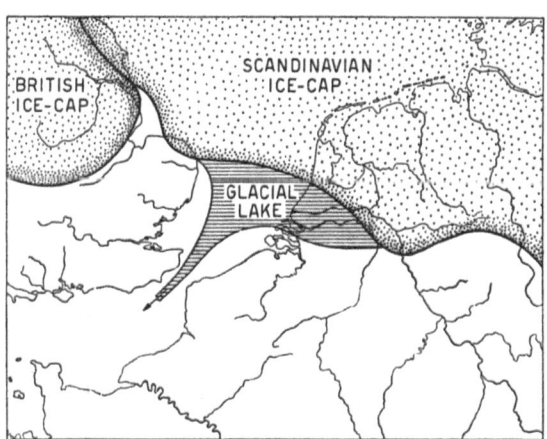

Fig. 25. Glacial lake of melt water formed in front of the ice-caps during the third or so-called Riss glaciation (After P. Tesch).

Another point of importance is that a large quantity of sand was deposited on the floor of the North Sea in the course of the advance and retreat of the continental ice-mass. The northeastern part of the petrographical map of the North Sea is occupied by what is called the A-group (fig. 26). The mineralogical composition of sands belonging to the A-group points to their northern origin. Apparently it is material transported by the Scandinavian ice-caps. They are glacial and fluvio-glacial deposits. In the same way the E-group, indicated by the vertical notation, points to English or Scottish origin. In a wide arc the Dutch coast is surrounded by sands of the so-called H-group, which consists of material predominantly from the Rhine.

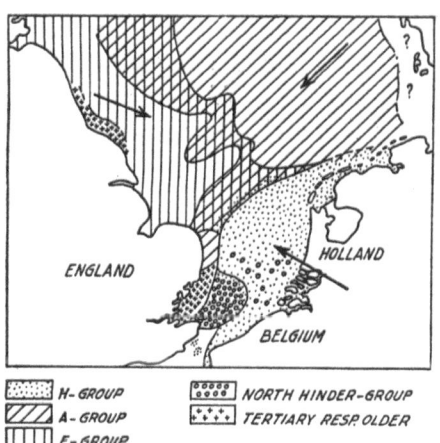

H-GROUP    NORTH HINDER-GROUP
A-GROUP    TERTIARY RESP. OLDER
E-GROUP

Fig. 26. Petrographical map of the North Sea (After J. A. Baak). Explanation in the text.

During the recession of the ice the melt-water drained into the oceans. Consequently, the sea-level rose and then the sand masses were cleared, sorted and re-arranged by the sea in the course of its advance.

It was in the course of the last advance of the sea that the present coast of Holland originated.

With the advance of the sea sand-masses originally transported and deposited mainly by rivers flowing from the south-east, accumulated in sand barriers to form the first coast-line of Holland. Hence, the coast is not a sand-bar or spit which grew north-eastward from near Calais, but it formed as the result of the combined action of waves and currents on superabundant sand masses on a gently sloping sea-floor. The sand barrier formed outside the original true shore-line. The greatest extent of the sea is shown by the most eastern extent of the old sea-clay which was deposited behind the sand barriers (fig. 27). Although accumulation of material in a direction at right angles to the coast predominated, a sand barrier- or lido-coast can never be formed without the additional agency of drifting along and parallel to the coast. And there seems not the slightest doubt that some of the sand barriers found "fixed" points in elevated and resistant remnants of glacial deposits. One of these was the Pleistocene of Wieringen. Another "skeleton" point was the low hill of glacial boulder clay on the present island Texel (fig. 27). The Dutch coast may be called "a lido or barrier coast possessing some liman qualities".

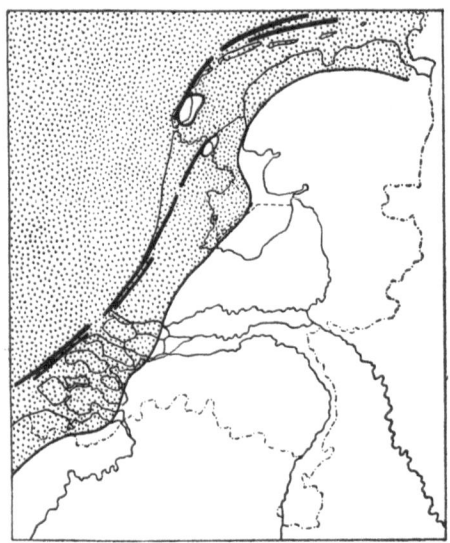

Fig. 27. Formation of the first sand barriers (After F. J. Faber).

*Movements of the land as well as of the sea-level.*

We know that about 3 metres of old sea-clay and some 2.5 metres of peat on top of it accumulated during a span of time which can be estimated at about 6000 years.

Obviously, however, the peat has grown under land conditions. Now, there was not the slightest chance for land conditions behind the sand barriers inasmuch as the barriers had several gaps through which the sea had free access. Of course, there must have been gaps from the very beginning even if there were no other reason than the necessity of an outlet for the water from the great rivers. And we know that such a gap is widened and scoured out by the sea.

The peat beds give the most convincing evidence that their formation was made possible by a retreat, I mean by a relative fall of sea-level.

The graph has to be altered accordingly (fig. 28). As we do not know what happened exactly, we shall suppose a stable land surface and then draw a graph for the relative movement of sea-level. Such a procedure implies the acceptance of a subsequent relative rise of the sea in more

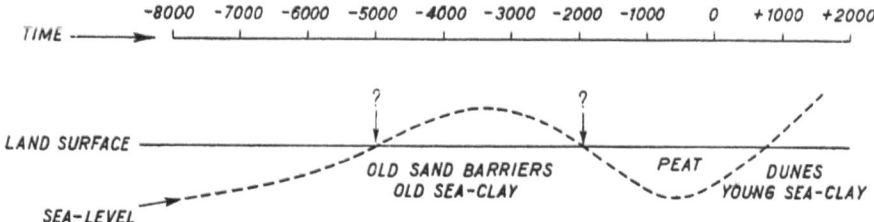

Fig. 28. Graph showing the relative movement of sea-level when accepting a stable land surface.

recent times. For, without artificial human construction present sea-level would extend twice as far eastward as it extended during the formation of the old sea-clay (fig. 19).

Now we may correct the term relative fall of sea-level and say that it was a true lowering of sea level.

As early as 1919 Daly came to suspect a sub-recent world-wide lowering of sea-level by a few metres.

I shall mention only a few examples.

A wave-cut bench having an elevation of some 6 metres above present mean sea-level is known from numerous islands in the Pacific.

Coral reefs are excellent tide-gauges. No wonder that many examples are known of reefs that have emerged by the same amount. Plate II, A portrays such a reef in the bay of Batavia. A very remarkable situation along the south coast of Java is shown in fig. 29. The surf breaks against a gently sloping limestone plateau, which is an emerged coral reef. Due to erosion its surface now lies less than 6 metres above mean sea-level. Sea-water is thrown on the limestone plateau by the high breakers of the Indian Ocean. It is hampered in its course by a ridge consisting of a fossil dune formation. In one spot the sea-water finds its way through an erosion gap in the old ridge. In this spot a phenomenon is displayed which probably is unique in the world, viz. a waterfall of sea-water discharging into a river, which a mile further on debouches into the Indian Ocean (Plate II, B).

The amount of 6 metres is again found along the shores of South-west and also East-Africa and Cape Town. Traces of a 6 metre shore-line have been mapped in South Africa along a distance of 1500 kilometres.

Again the same uniformity of emergence was found on islands in the central part of the southern Atlantic. Saint Helena is an example. In short, the numerous examples show that the theory of a sub-recent fall of sea-level seems firmly established. It has been strengthened by an ever growing mass of evidence from all over the world.

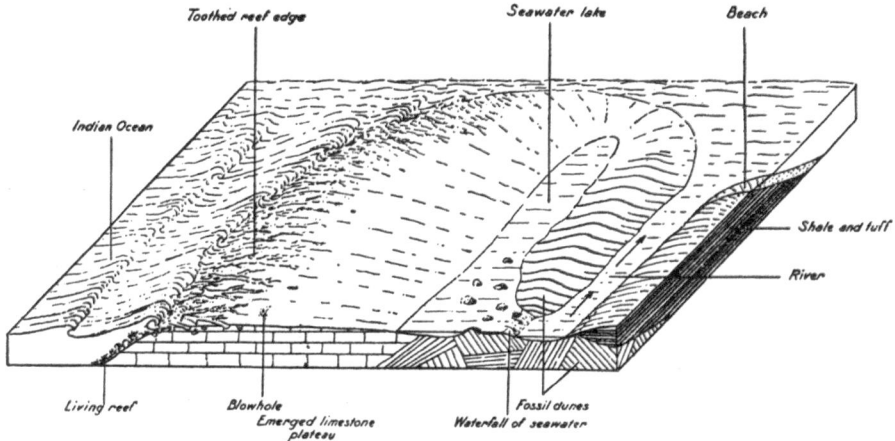

Fig. 29. Block-diagram explaining Plate II (B).

Of course, no emergence of a coastal tract can be expected in areas of notable subsidence like Holland. But Holland would look very different if no sub-recent lowering of sea-level had taken place; and certainly there would be no peat deposits.

As a matter of fact, the amount of the emergence shown by more stable parts of the earth is of paramount importance for untangling the interwoven effects of sea-level movement and land movement in Holland. Possibly sea-level actually dropped by an unknown amount say $6 + x$ metres and rose subsequently, but in any case it is now 6 metres lower than when the recession started. Possibly the $6 + x$ metres fall of sea-level was caused by a temporary growth of the ice-caps, some four thousand years ago.

Obviously, the amount of 6 metres has been surpassed by subsidence of the bottom along the Dutch coast. And, evidently, the effect of the fall of sea-level ceased at about A.D. 850 [1]), when a new advance of the sea

---

[1]) This figure seams the right date for the two western provinces of Holland. It may be different in other districts more to the north or the south.

Plate II

*(A)* Emerged coral reef in the bay of Batavia.

*(B)* A waterfall of sea-water discharging in a river, Tji Laut Eureun, south coast of Java.

started along the Dutch coast. Hence, the downward movement of the land must have been in the order of 6 metres in about 2850 years which means an average sub-sidence of the land of 21 centimetres per century.

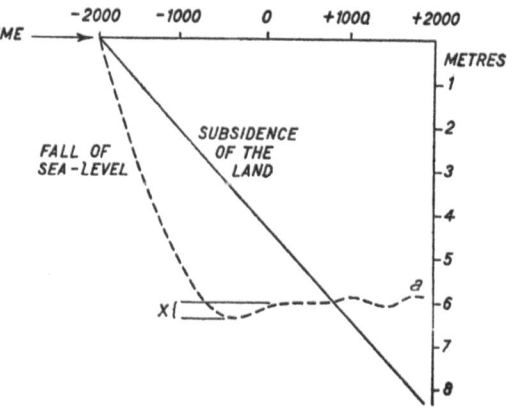

The full-drawn line (fig. 30) shows the supposed subsidence of the land at a constant rate during the last 4000 years. The dot-dash line represents the fall of sea-level by an amount of 6 metres. At the point where the two curves intersect, the most recent relative rise of sea-level starts.

Fig. 30. Graph showing the combined effect of subsidence of the bottom and the 6 m fall of sea-level.

The dotted line shows a small undulation at the extreme right (a). For, it was found from the records of 71 tide gauges over the whole world that since about A.D. 1890 sea-level has risen at a secular rate of about 12 cm. It is well known that the glaciers and ice-caps have shrunk all over the world in the course of the last century especially since about 1890 and this apparently is the cause of the rise of sea-level during the last decades.

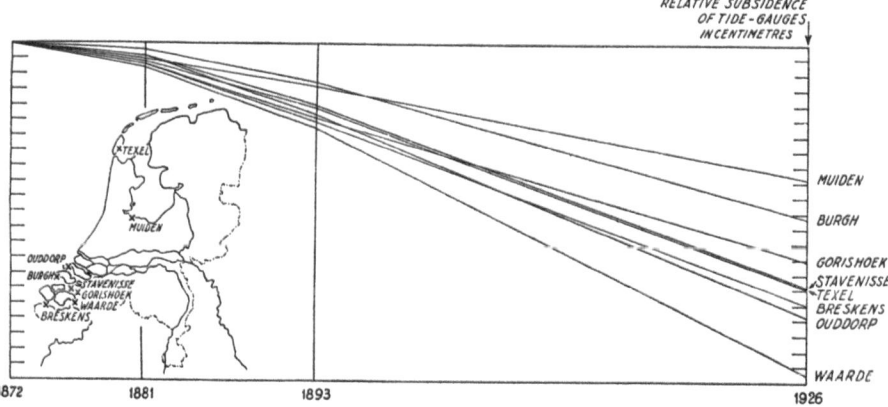

Fig. 31. Relative movement of sea-level as registrated by 8 tide-gauges between the years 1872 and 1926 (Adapted from B. G. Escher).

Now let us compare this result with data recorded by tide-gauges in the coastal district of Holland. If the amount of 12 cm. be added to the 21 cm. subsidence of the land, tide-gauges in the western coastal districts of Holland ought to show a relative rise of sea-level of 33 cm. Now, 30 cm. is the amount calculated by several authors from tide-gauge data. The diagram of fig. 31 illustrates this phenomenon. It shows moreover the increasing movement of later times which is obviously due to an additional eustatic rise since A.D. 1890.

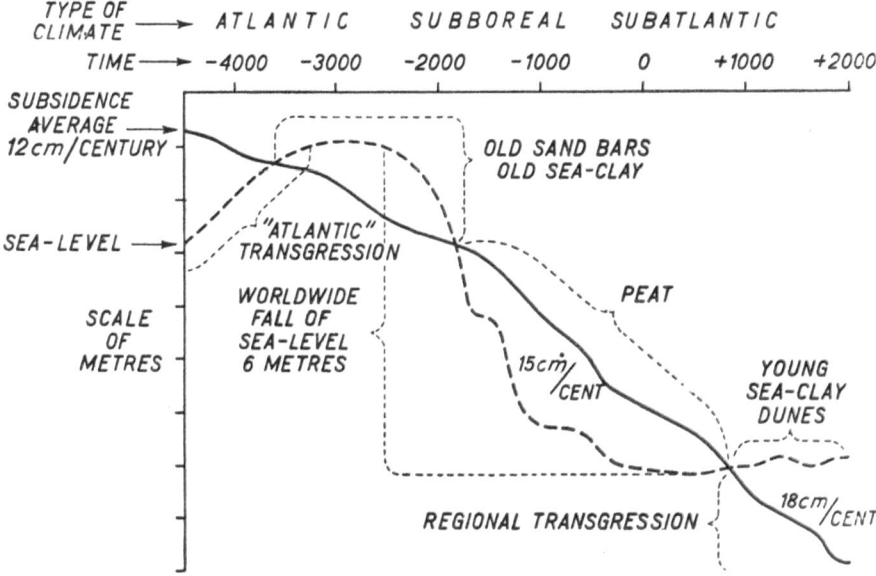

Fig. 32. Graphic representation of the supposed movements of sea-level and subsidence of the bottom.

Notwithstanding the remarkable agreement with the records of tide-gauges the graph of fig. 30 has to be regarded as a rough approximation. For, there are several uncertain factors. One is the preliminary character of the time-scale, the starting point 2000 B.C. being only approximately at the right place. Another uncertain factor is the exact amount of the fall of sea-level. Then, the subsidence of the land is represented by a straight line. It may be that it should be replaced by an undulating line showing one or more accelerations and retardations. Finally, the records from tide-gauges have to be taken *cum grano salis*. These and several other factors, which will be left out of discussion here, were taken into account in the construction of fig. 32.

I have tried to make clear that the historical succession of deposits in the western part of Holland can be explained only by the combined

movements of the land as well as of the sea-level. But there is a second
problem which is explained by the interaction of these movements. For,
now the destruction of the old sand barriers and the reasons for the
formation of the dunes are no longer a puzzle. Since Roman times the
coast has receded in many places, for example two kilometres near Den
Helder between 1571 and 1866 A.D.

The shape of the coast adjusted itself to the relative rise of sea-level and
the prevailing system of currents. For this reason the present line of the
coast dissects the trend of the former coast and the old sand barriers.

The sea received a great quantity of sand not only from the demolished
sand barriers, but also from deeper parts due to the adjustment of the
submarine profile to the new coast line (the stippled as well as the vertically

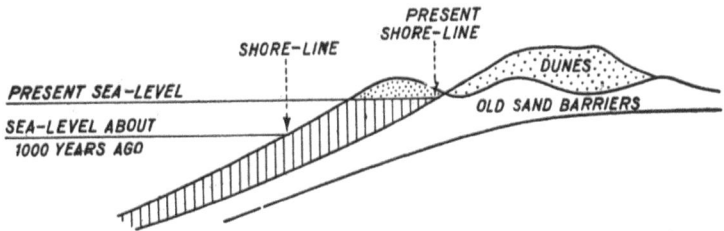

Fig. 33. Formation of the dunes due to a relative rise of sea-level.

shaded part of fig. 33). This material was heaped up in the younger dunes.
As a result of prevailing strong westerly winds the younger dunes are
formations which are very different from the low and parallel sand-
barriers (the so-called "old dunes") which were predominantly modelled
by the action of the sea.

As another result of the relative rise of sea-level many invasions of the
sea occurred since the ninth century (leaving the younger sea-clay
deposits behind). At this stage man had to occupy himself intensively
with the situation. Without his strenuous efforts Holland would have
become a prey of the waves. In the course of many centuries large stretches
of land have been added to the country in an eternal struggle with the sea
(fig. 34).

There is indeed a germ of truth in the saying: "God created the world
with the exception of Holland which was gained by the Dutch themselves". And the people are still busy reclaiming new fertile soil for future
generations. With that aim the great dyke was built between Holland and
Friesland. It was completed in May 1932. Five years later the salt water of
the former Zuiderzee had become a freshwater body, the Ysel-lake. Two
new polders have already been reclaimed from this lake, one in the west

— the Wieringermeer polder — and one in the east — the North-East polder which was finished in 1941. New ones will be added in the future.

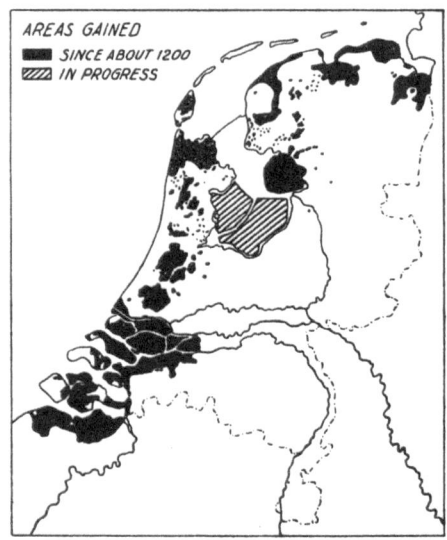

Fig. 34 Reclamation of land since about A.D. 1200 and future plans of gaining further parts of the Ysel Lake (After J. van Veen).

But the older polders of Friesland, Holland and Zealand as well as the coast need constant supervision. It is in these regions, which are situated a few metres below sea-level that towns like Amsterdam and Rotterdam have been built.

Subsidence of the bottom is our great enemy. Due to the fundamental action of that factor the sea has become automatically its ally, an ally of ever increasing danger. And since we cannot stop our arch-enemy deep beneath, we fight his most dangerous ally which tries to invade the coastal provinces. There are only two things we can hope for the future. One is a growth of the ice-caps — not too much — and a consequent lowering of sea-level, the other that the bottom movement will slacken its pace considerably or, even better, stop entirely — the sooner the better.

### THE CAUSE OF THE SUBSIDENCE OF THE LAND SURFACE

I hope that I have succeeded in making clear why Holland is a flat country bordered by a strip of dunes and why the surface of the western provinces is a few metres below sea-level.

Probably, however, you want to know what the cause is of the subsiding movement. I have mentioned already that the Netherlands form part of the North Sea basin. A basin is a saucer-shaped depression in the earth's crust, formed by a gradual subsidence of the bottom. Generally the depression is filled up with sediments while its floor subsides.

The structural boundaries of the North Sea basin consist of very old lineaments or axes of elevation. In the east is the so-called axis of Erkelenz. The southern boundary is formed by the massif of Brabant; the present morphology of Belgium — e.g. the pattern of the rivers — still reveals the influence of this important element; the stream pattern is nicely adapted to it.

South of the Brabant massif is another saucer-shaped depression in the earth's crust: the basin of Paris. The southern part of the Brabant massif shows an inclination towards France, the northern part towards the Netherlands. Near Maastricht the massif dips suddenly to great depth. Eastward follows the faulted area of the Limburgian coal district (fig. 35).

It will be clear that in order to see something of older formations we have to proceed towards the eastern and southern boundaries of the basin. The boundaries between several horizons of the Tertiary are more or less parallel to these structural boundaries, which remarkably enough roughly coincide with the eastern and southern frontiers of the Netherlands.

The origin and subsequent subsidence of the floor of the North Sea basin date from a very long time ago.

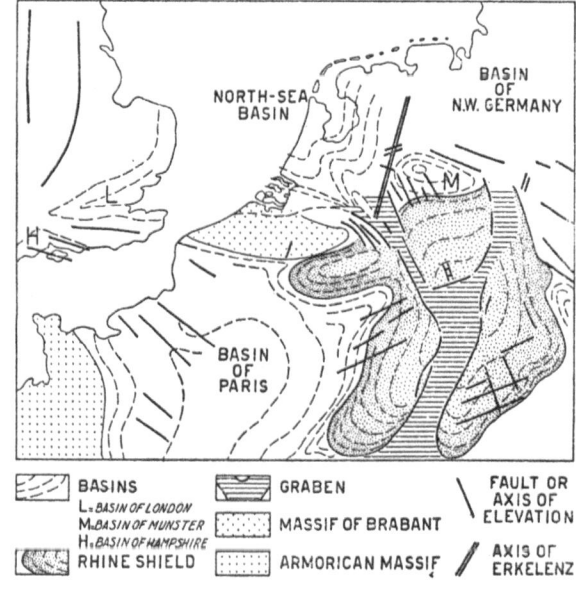

Fig. 35. Structural boundaries of the south-eastern part of the North Sea basin.

They started some 220 million years ago, in a period which is called the Permian and the submergence has continued up to our own days (fig. 36).

The downward movement of the land-surface is the result of the interplay of several factors. One possible factor is compaction of the sediments. It is quite conceivable that the sub-recent sediments are only slightly consolidated. Doubtless this is true of the peat deposits. Perhaps you know that all the houses and greater buildings of big Dutch towns like Amsterdam are built on a foundation of large wooden piles which have been rammed into the bottom. Without such a precaution the buildings would sink down into the mud and peat beneath. And even now some of them are found to subside slowly. Compaction of the deposits, more especially of the peat layer, is perhaps one of the causes of the subsidence of the land-surface. But we do not have accurate data at our disposal which would allow an estimate of its amount.

What we know is that the North Sea basin is filled up with sediments ranging from Permian to recent times.

The total sequence of sediments that accumulated in the central part of the basin may be estimated at 7,500–9,000 metres. This means an

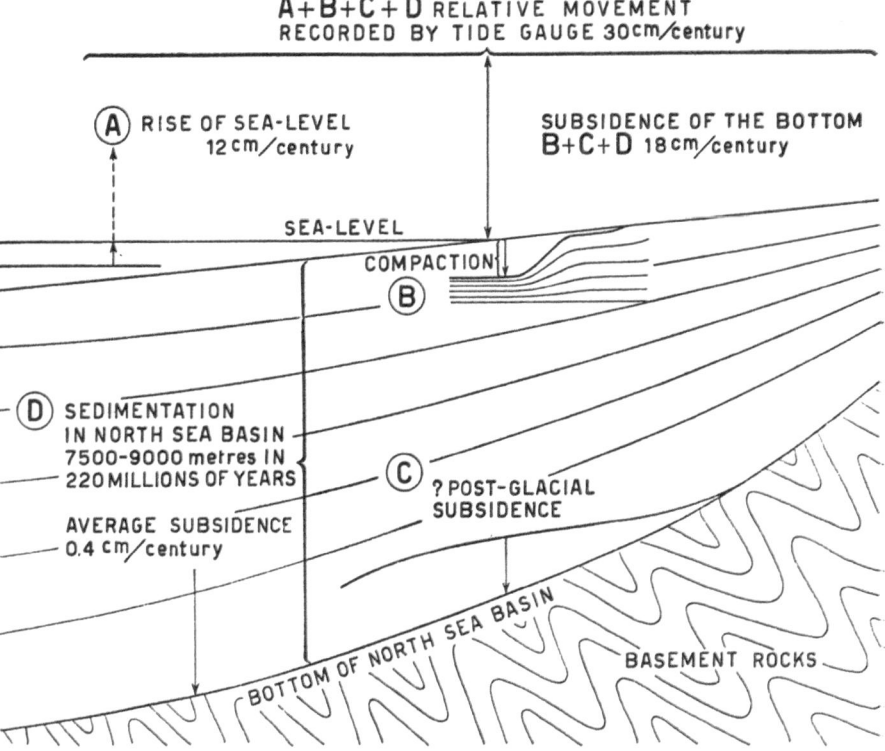

Fig. 36. Schematic section through the eastern part of the North Sea basin showing factors. discussed in the text.

average subsidence of o.4 cm. per century during the last 220 million years.

Hence, the subsiding movement of our own days is about 50 times the average rate of subsidence since Permian times.

One factor causing this abnormally high rate of movement may be compaction. If so, it is only a superficial phenomenon which has to be considered apart from a possible sinking of the basin as a whole. It has nothing to do with the sinking of the floor of the basin. It is a well-known fact in Holland that the more the land is drained the faster is the sub-sidence due to compaction of the peat layers.

Possibly a second factor is acting at a deeper level. To make this clear I shall have to turn to glacial phenomena for a moment. Scandinavia was basined under the load of the last or so-called Würm ice-cap. Probably the basined area was accompanied by a slight uplift of the surrounding belt, due to horizontal displacement of material below the crust. Of course the vertical scale of fig. 37 is strongly exaggerated. When the ice-cap melts the crust is unloaded and the floor

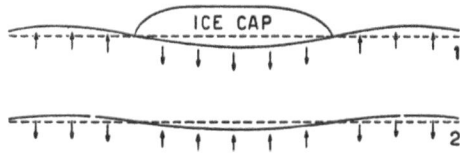

Fig. 37. Bottom movement caused by the growth and melting of an ice-cap.

will rise to re-establish equilibrium. On the other hand the peripheral zone, which was superelevated, will tend to sink down.

Now, the rising of Scandinavia is recorded by tide-gauges along the shore of the Baltic and Bothnian Gulf. It amounts to as much as 1 m. per century in the central area of the former ice-cap.

The zero-line (or hinge line) between the two opposed movements runs over Denmark. To the south of the hinge-line the land surface may be expected to sink. Possibly part of the Netherlands belongs to the downward moving circumferential belt of the former ice-cap. This is no more, however, than a theoretical suggestion. No data are available to substantiate the theory and of course nothing can be said of the rate of subsidence due to that hypothetical factor.

Possibly, however, a third factor has been of influence (see fig. 36). Problematic processes beneath the earth's crust caused the initiating of the downward movement of the basin. The floor continued to subside at an average rate of 0.4 cm. per century. But undoubtedly periods of accelerated and retarded subsidence alternated. You will notice presently that accelerated movements appear to be correlated in point of time with epochs of decreasing compression in the earth's crust. A time of strong mountain-building like the one we are living in now is such an epoch. It is the time when the folded and crumpled lighter material of the upper part of the crust gets an opportunity to rise. Sub-crustal material flows towards the area beneath the rising mountain-belts. At the same time the sinking movement of basins is accelerated. Probably there exists a causal relation between the two phenomena. These few words may suffice to point to the possibility that sub-crustal processes which are world-wide are causing an accelerated subsidence of the North Sea basin in our days. Again, however, no data are available on which a quantitative estimate of the influence of that factor could be based. A problem for the future is to analyse the curve of fig. 32 into three or more separate curves.

All we can determine for the moment is the approximate amount of the combined effect of the three factors considered.

In itself, however, the subsiding movement shows some very characteristic features which indicate that sinking of the bottom is an important factor. Both the occurrence of tilting as well as the intermittent character

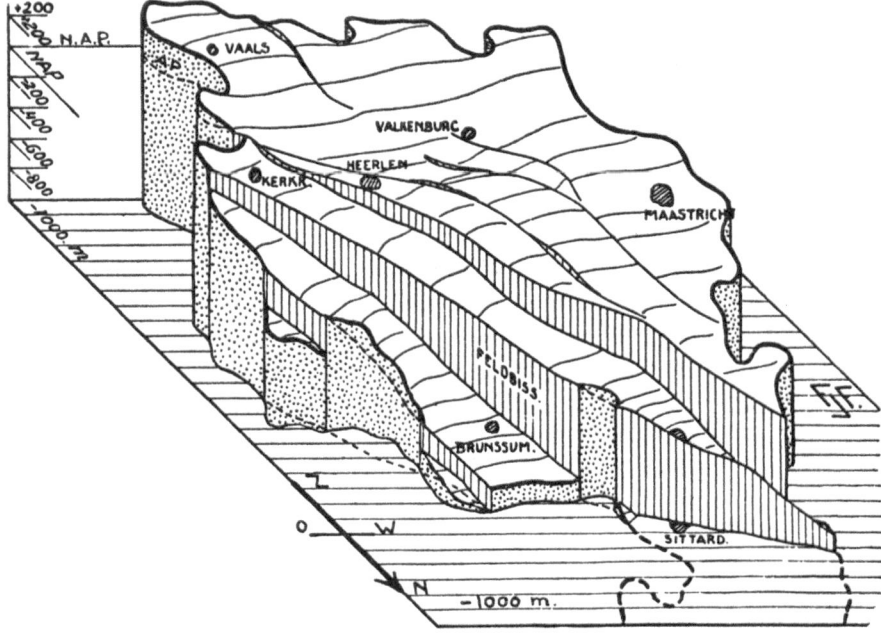

Fig. 38. Surface of the Carboniferous in the sub-soil of the southern Netherlands, showing pattern of faults and north-westward dip. Thickness of overlying Mesozoic and Tertiary strata increases in the same direction. (After F. J. Faber).

of the movements can be clearly demonstrated. Moreover, this examination will show us that these movements are intimately related to other phenomena.

To complete our picture we must consider another structural element in the southeast which has been left out the of discussion so far, viz. the so-called Rhine Shield. The Vosges, the Schwarzwald, the "Schiefergebirge" and the Ardennes belong to it. Unlike a basin it forms a dome-shaped elevation. The roof of the shield collapsed and a three-fold system of graben came into existence viz. the Rhine-graben, a north-eastern graben and the north-western graben-system of the Netherlands (fig. 35). The first indication of the updoming movement of the Rhine Shield dates from post-Variscian times. Repeatedly the movement was rejuvenated. In the Oligocene the Rhine Shield and the graben underwent a very marked rejuvenation. This was also the time of very strong movements in the Alps.

The northwestern branch of the graben system enters the Netherlands via southern Limburg.

The block-diagram (fig. 38) shows the whole system of faults and graben to dip in a northwestern direction. The thickness of the over lying strata increases towards the northwest which is towards the centre of the basin. It is in the same direction, i.e. towards the basin, that the

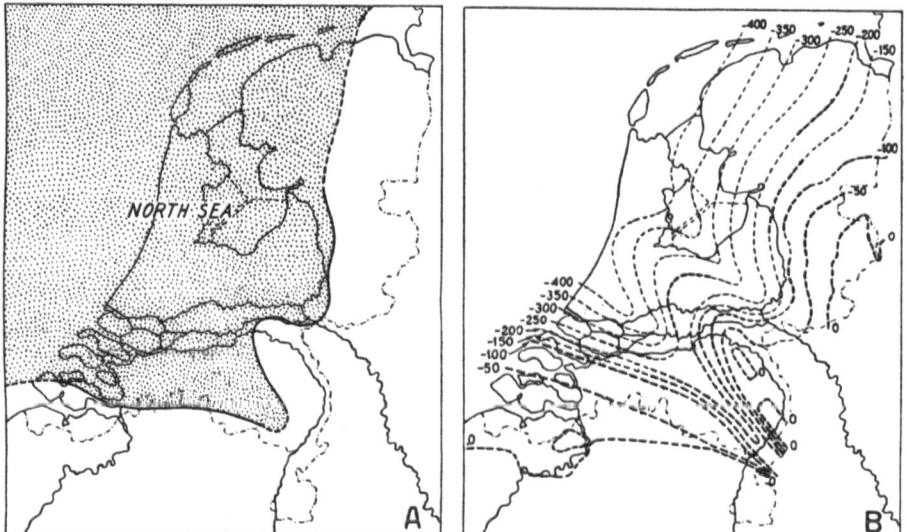

Fig. 39. The graben in the Netherlands as displayed by (A) deposits of the first ice age, (B) the pre-Pleistocene surface (After P. Tesch).

fault system broadens to form what is called the central graben of the Netherlands (fig. 39).

Repeatedly movements took place along the faults and the differential movements along the blocks had a great influence on the distribution and varying thickness of the sediments in the basin. One will notice the effect in maps which show the surface of any particular stratum in the basin. The first picture shows a map of deposits belonging to the first ice-age (fig. 39, A). The boundary between marine and terrestrial deposits clearly reveals the influence of two graben-systems.

The second shows isobases drawn on the pre-Pleistocene surface (fig. 39, B); again the same influence is clearly demonstrated. Moreover, one can note once more the influence of the stable structural boundaries in the east and south.

Differential movements caused the big graben to become subdivided by several horst-like structures. The Peel-horst is a well-known example.

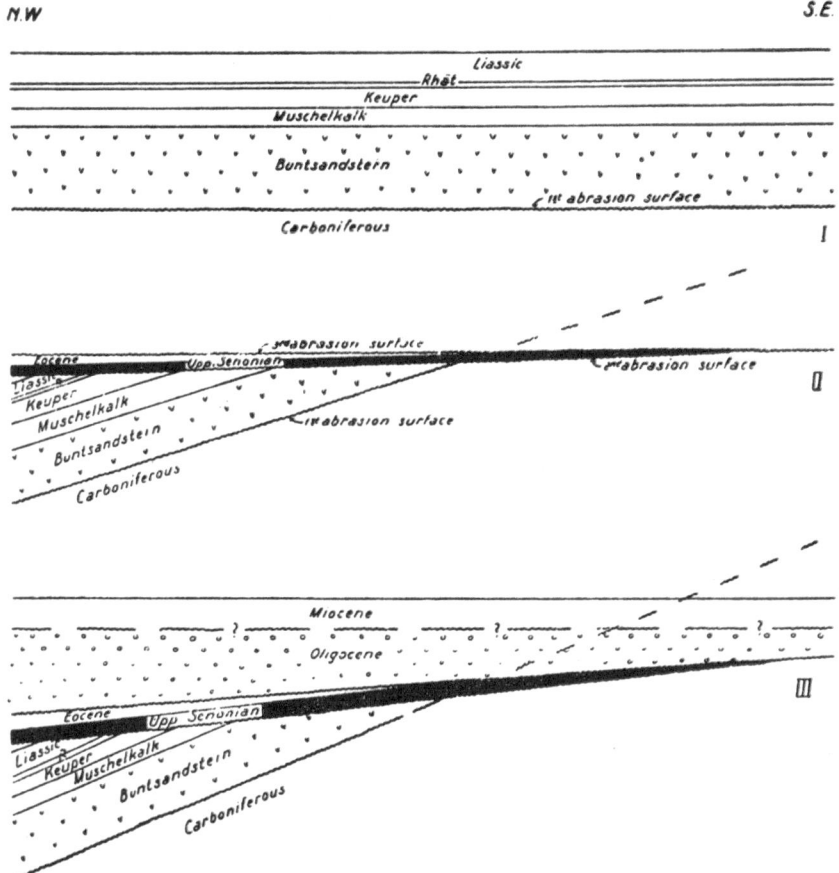

Fig. 40. Intermittent tilting of the bottom towards the basin, as shown by Mesozoic and Tertiary deposits (After L.U. de Sitter).

Several blocks at a comparatively high level were revealed by gravimetric explorations carried out during recent years.

But now let us return to the subsiding movement of the basin in general. The following remarks demonstrate very clearly that the downward movement was accelerated at certain epochs. Moreover, you will notice that the movement was a sort of tilting process in which the structural boundaries in the east and south formed the hinge-lines.

From available data a tilting of the bottom along a hinge-line in the south may be deduced for two epochs, viz. (1) post-Lias though pre-Senonian, (2) Oligocene (fig. 40).

After horizontal deposition of the pre-Senonian sequence of strata a tilting movement took place. Then, there followed a period of rest during which Senonian and Eocene sediments accumulated. Then again the whole sequence was tilted so as to cause a renewed dip towards the north-west.

The pre-Senonian warping is correlated, in point of time, with the Upper Cimmerian epoch of movement known from many geosynclines all over the world. The Oligocene movement was synchronous with one of the principal phases of folding of the Alps. Moreover, both the epochs of movement are marked by major phases of movement along faults of the southern Netherlands, by the formation of the Rhine graben, etc.

A still more recent submergence of the North Sea basin is clearly

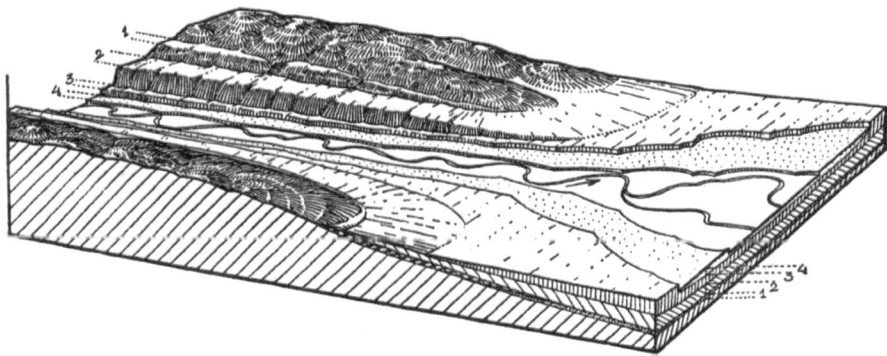

Fig. 41. Subsidence of the North Sea basin as revealed by Pleistocene river deposits. 1, 2, 3 and 4 indicate deposits decreasing in age. To the left the oldest deposit (1) is higher than (2), and (2) is at a higher level than (3) and so on; to the right (4) is resting on top of (3) and (3) on top of 2 etc. The last named series is formed where the bottom subsided. The sequence of terraces, to the left, was formed under intermittent rise. (After K. Oestreich).

revealed by the distribution of surface deposits. In the subsiding basin the gravels of the Rhine and Maas have accumulated in the usual way, i.e. the more recent sediments rest on top of the older deposits (fig. 41). However, if one proceeds along the rivers towards the east or south one will meet the well-known sequence of fluviatile terraces that formed along the incised channels of the streams. Proceeding from the highest terrace to the lowest the age of the deposits diminishes (from Lower Pleistocene, or even Upper Pliocene, to sub-recent). The result of these circumstances is the remarkable fact that the terraces cross each other if seen in a longitudinal profile of the stream. The point of crossing is the hinge-line between the subsiding basin on one side and the rising movement of an adjacent structural element on the other side.

It is one of the most striking results of an examination of basins as well as of dome-shaped elevations that their time of origin and rejuvenation is related to certain epochs of folding and mountain-building, though they sometimes originate at a great distance from such zones. The subsiding movement of basins begins after a special epoch of folding. And in addition to this, certain phenomena in the history of these basins, such as unconformities, faulting, faint folding and a repeated tendency to subside appear to be related, in point of time, to specific phases known in folded chains. One thing and another not only point to a deep-seated terrestrial process, but also show that this process has a world-wide activity. The greater part of the basins (as well as of the troughs) formed after certain Variscian phases of movement. A similar and exceptionally intense formation of basins succeeded certain Alpine phases. Those basins which are connected with Caledonian epochs, as far as their origin is concerned, are undoubtedly far less numerous, and only a few can be associated with Lower and Upper Cimmerian phases. Perhaps this can be more clearly demonstrated in a graphic representation (fig. 6, 8, and 18).

It certainly is not a mere accident that the intensity of mountain-building is related in point of time to the great peaks of the formation of basins. Epochs of folding, i.e. periods of increasing pressure in the earth's crust, are succeeded by periods of decreasing compression, which also represent periods of mountain-building. As soon as the activity of a period of intensive compression ceases, the mountains not only begin to rise, but the negative elements, too — the basins — begin to take shape.

The time of origin of the North Sea basin as such, especially of its eastern and western boundaries, coincides with the Saalian phase. The formation of the basin of Paris dates from the Lower Cimmerian phase, but I will not go into more details.

I hope that I have succeeded in showing that the origin of basins is due to deep-seated processes beneath the earth's crust, and particularly that periods of decreasing compression in the earth's crust are characterized both by a rising movement of the folded chains, and the beginning of local tendencies to subside, — in other words the beginning of the formation of basins. So, the whole history of the North Sea basin is only one small element in the great *Symphony of the Earth*.

## REFERENCES

"A country below sea-level" was the topic of several addresses on various occasions and in different countries (Cambridge, Delft, The Hague and Amsterdam in 1946, Copenhagen Stockholm and Oslo in 1947, Liège and Utrecht in 1948, Ghent in 1949).

Publications worked up in the previous text are listed in the following bibliography.

AHLMAN, H. W.SON., *The present climatic fluctuation.* The Geogr. Journal CXII 1949, pp. 165–196.

BAAK, J. A., *Regional petrology of the southern North Sea.* (Academ. Thesis Leyden, 1936).

BLANCHARD, R., *L'Origine des Moëres de la plaine maritime de Flandre.* (Bull. Soc. Geogr. 31, pp. 337–346, 1917).

BRIQUET, A., *Le littoral du Nord de la France et son évolution morphologique.* (A. Colin, Paris 1930).

DALY, R. A., *The changing world of the Ice Age.* (Yale Univ. Press 1934).

DUBOIS, EUG. *Over het ontstaan en de geologische geschiedenis van vennen, venen en zeeduinen.* (Arch. du Musée Teyler, ser. 3, vol. 4, 1949).

DUBOIS, G., *Classification du Quaternaire du Nord de la France et comparaison avec le Quaternaire Danois* (C.R.A. des Sci. 179, 1924.)

DUBOIS, G., *Recherches sur les terrains quaternaires du Nord de la France.* (Mem. Soc. géol. du Nord, 8, 1924).

ESCHER, B. G., *Het vraagstuk van de daling van den bodem van Nederland.* (Geologie en Mijnbouw 1940, pp. 173–196).

GODWIN, H., *Pollen analysis and Quaternary Geology.* (Proc. Geol. Assoc. LII, 1941, pp. 328–361).

GODWIN, H., *Coastal peat beds of the North Sea region as indices of land- and sea- level changes.* (New Phytologist 44, 1945).

GUTENBERG, B., *Changes of sea-level, post-glacial uplift and mobility of the earth's interior.* (Bull. Geol. Soc. America 52, 1941).

HALET, F., *Contribution à l'étude du Quaternaire de la plaine maritime Belge.* (Bull. Soc. Belge de Geolog. Pal. et Hydrol. XLI, 1931).

KUENEN, PH. H., *De zeespiegelrijzing der laatste decennia.* (Tijdschr. Kon. Nederl. Aardrijksk. Genootschap 62, 1945).

LORIÉ, J., *Binnenduinen en bodembewegingen.* (Tijdschr. Kon. Nederl. Aardrijksk. Genootschap 10, 1893).

POLAK, B., *Een onderzoek naar de botanische samenstelling van het Hollandsche veen.* (Academ. Thesis. Amsterdam 1939).

SITTER, L. U. DE, *The Alpine geological history of the S. Limburg coal district.* (Jaarverslag 1940–'41 Geolog. Bureau Mijngebied 1942).

STEERS, J. A., *The coastline of England and Wales,* (p. 175 and 426–439. Cambridge Univ. Press 1946).

STEVENS, CH., *Le relief de la Belgique.* (Mém. de l'Institut Geolog. de l'Université de Louvain 12, 1938).

TAVERNIER, J., *L'évolution de la plaine maritime belge.* (Bull. Soc. Belge de Geologie LVI, 1947, pp. 332–343).

TESCH, P., *Duinstudies I–XIII.* (Tijdschr. Kon. Nederl. Aardrijksk. Genootschap 27–38, 1920–1930).

TESCH, P., *De Noordzee.* (Meded. Rijks Geolog. Dienst A, no. 9, 1942).

TESCH, P., *Nederland bij het begin van onze tijdrekening.* (Tijdschr. Kon. Nederl. Aardrijksk. Genootschap LXI, 1944, p. 456–458).

TESCH, P. en REINHOLD, TH., *De bodem van het zuidelijk uiteinde der Noordzee.* (Tijdschr. Kon. Nederl. Aardrijksk. Genootschap LXIII, 1946, pp. 72–84).

UMBGROVE, J. H. F., *Periodical events in the North Sea basin.* (Geolog. Magazine 82, 1945, pp. 237–244).

UMBGROVE, J. H. F., *Origin of the Dutch Coast.* (Proceed. Kon. Nederl. Akad. v. Wetenschappen L, 1947, pp. 227–236).

VEEN, J. VAN, *Onderzoekingen in de hoofden.* (Acad. Thesis, Leyden 1936).

VEEN, J. VAN, *Dredge, drain, reclaim.* (Nijhoff, The Hague, 1948).

## ACROSS THE SWISS ALPS

### INTRODUCTION

The question how a folded mountain-chain came into being is some-times answered by a very simple demonstration *ad oculos*. A napkin or a tablecloth is pushed into folds over the smooth surface of the table on which it was spread. For most types of mountain-chains (as for example that of the Alps) this apparently instructive experiment is not to the point. But it is often held to be relevant with regard to the Jura Mountains. However, when examining the available evidence it will appear that Nature's way of rumpling the tablecloth of the Juras was essentially different from our usual attempt at imitation. In order to make this point clear it will be unavoidable to take into consideration the wider surround-ings of the Jura Mountains. In fact, we shall have to cross the entire chain of the Alps from north to south. When doing so the problem of the Juras will gradually become cleared up, though not fully before we shall have reached the end of our trip.

When travelling through Switzerland from north to south, from the Rhine valley at Basle to the southern Alps of Lugano, one will notice how very different the scenery is in various parts. First the Jura Mountains are crossed, an arcuate bundle of more or less parallel ridges, which are only moderately high. At Olten or at Solothurn the Swiss plain is reached, extending southward over a distance of some 20 miles. Suppose we take the Gothard route and arrive at the Lake of Lucerne. It is immediately apparent that the Rigi is a type of mountain differing greatly from the High Calcareous Alps which build up the attractive scenery south of the lake. As a matter of fact, the differences in scenery are the expression of differences in rocks and structure (cf. fig. 42). The Rigi, for example, is a tilted and raised block of conglomerates and sandstones. Large stretches of the Swiss plain consist of similar material but the beds are not so disturbed as in the Rigi. On the other hand the High Calcareous Alps consist of huge folds and of slices of limestone and other strata. These were originally deposited in a trough situated much farther south. The contents of the trough was strongly compressed and displaced in a northerly direction to become the so-called Helvetian nappes of the present High Calcareous Alps. We cross them along the famous Axenstrasse and after having made some trips eastward and westward, arrive at the Aar- and Gothard massifs where the landscape is dominated by very old crystalline schists and granites with their typical erosion features. A

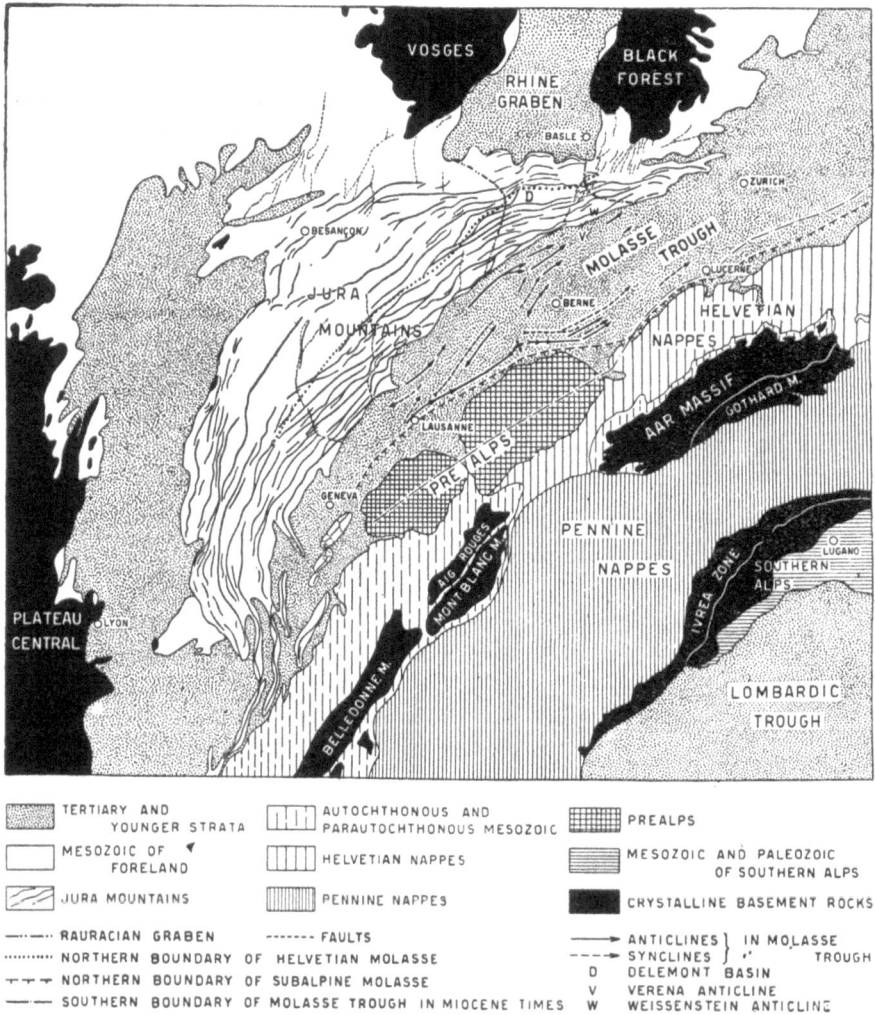

Fig. 42. Major structural units of Switzerland.

similar region is the Aiguilles Rouges–Mont Blanc massif. Descending from the Gothard along the Val Tremola the route leads to Airolo on the boundary between Gothard and the northernmost chains of the Pennine Alps.

In order to study the enormous recumbent folds which make up the structural pattern of the Pennine Alps an excursion to the surroundings of Zermatt is the next point on the programme. Thence we shall pass the Ivrea zone, the surroundings of Bellinzona, the southern Alps of Lugano and their eastward continuation in the Orobic zone and the Lombardic Alps. The latter are on Italian territory where our excursion ends in the plain of Lombardy. However, the more trained mountaineers will not have earned their rest at Milano until they have made a final excursion to the plutonic bodies of Bergell and Adamello.

The only purpose of this geological trip across the Swiss Alps is to gain some insight as to its structural features and their bearing on the problems of the origin of the Jura Mountains. The stratigraphic side which — of course — forms the foundation on which the unravelling of the structural history of a region has to be based will be left out of consideration as much as possible in our short account of these majestic mountains.

Similarly we cannot enter into the historical aspects of the famous explorations carried out by numerous scientists in the Swiss Alps. Let us only mention that among the great mountain systems of the world no other region has been explored by so many excellent geologists or in such great detail.

More than a century of geological mapping and thinking found its climax in the work of Argand, whose brilliant synthesis of the Alps published in 1916, has been generally accepted. In the light of still later discoveries some of the most fundamental aspects of his theory can no longer be maintained and — as I hope to make clear — have to be replaced by a quite different concept.

A second point of importance is that the new conception is supported and amplified by results of geophysical investigations. But that is a more difficult topic, which will not be taken up until Chapter V.

THE JURA MOUNTAINS

The Jura Mountains consist of an arcuate bundle of fairly regular parallel ridges.

A characteristic feature of the scenery is the general congruence between landscape and internal structure. Generally a hill corresponds to an upward fold or *anticline* whereas a valley corresponds to a downward fold or *syncline* (fig. 43). A few rivers, however, have cut right across the general trend of the folds. The Birs river is an example. The coincidence between the ridge and vault-like or "anticlinal" structure of its strata can

Fig. 43. Type of Jura folds showing congruence between structure and topographic relief as revealed in the "cluse" of the Birs river, near Moutier. View from the Graitery anticline.

be seen very clearly where the river has cut its valley through the Raimeux range near Moutier (fig. 43). Albert Heim compared the Jura Mountains with a tablecloth that, being pushed to one side, has been rumpled into ridges and valleys. The Birs was already flowing in a roughly south-north direction before the "tablecloth" became rumpled. And during the arching up of the crests it gradually incised its valley deeper and deeper right across the rising anticlines.

A very valuable source of information in exploring the structure of the Swiss mountains was provided by the many tunnels. Structural prognoses could be checked and in several areas intricate structures encountered below the surface could be unravelled with great accuracy owing to the uninterrupted exposure along the walls of the tunnels. Fig. 44, B shows an example from the Jura Mountains. In this section detailed mapping of strata at the surface and in the small gorge of the Gsieg river were supplemented by data obtained during the construction of the Hauenstein tunnel. The northern ranges of the Jura have been pushed over the "fore-land" in the shape of clear-cut thrusts. Once the structural pattern of the mountains is known one can very clearly trace its main features in the scenery. Thus fig. 44 A is a view of the same region though slightly more

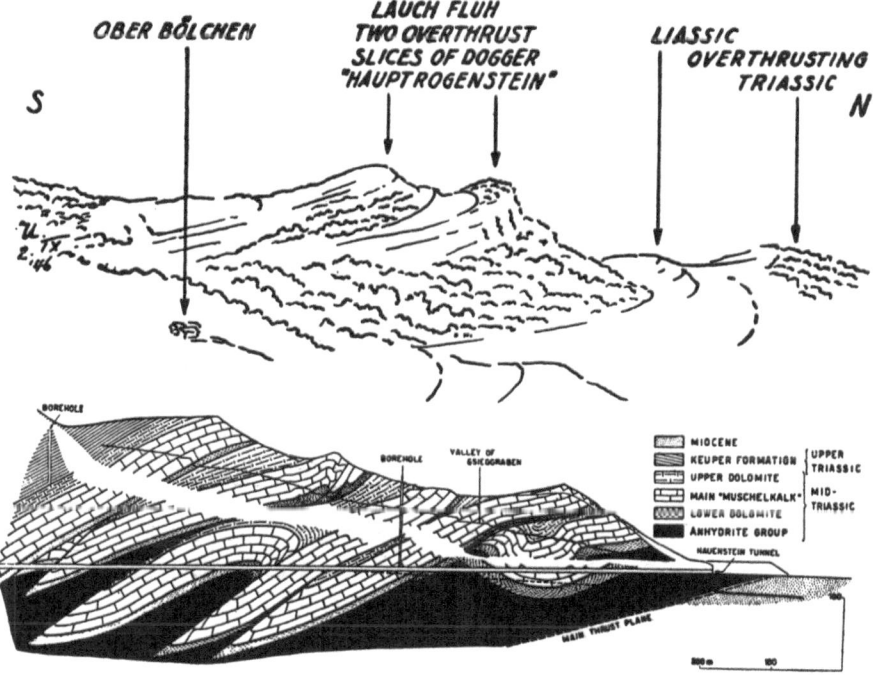

Fig. 44. (A) Overthrust structures featured in the scenery west of the Hauenstein tunnel.
(B) Overthrust structures of the northernmost Jura folds on their foreland in the area of the Hauenstein tunnel. (After D. H. Thornburg).

towards the west. Some of the overthrust slices stand out distinctly as features in the scenery.

If the Jura Mountains can be compared with a tablecloth that has been rumpled up and pushed to one side, where is the table and what is the equivalent of the hand pushing the cloth?

Detailed explorations showed the greater part of the Jura Mountains to be built up of strata belonging to a period which is called Jurassic. During Jurassic times a sea left extensive deposits in this area. These strata have obtained their name from that of the Jura Mountains where they are represented in a typical sequence.

Below the Jurassic strata older formations come into view. They are called Triassic.

We know that the Triassic rests on still older strata called Permian and these, in turn, in this region, rest on a very old complex of intensely crumpled and metamorphosed rocks, the crystalline schists.

We are pretty sure that the whole sequence of older rocks extends beneath the Juras, for southeast of Basle they are known to dip under the northernmost Jura folds which overthrust on their plateau-shaped foreland. In the Vosges and Black Forest the old schists occupy large areas at the surface.

It is highly remarkable, however, that the older rocks actually appear nowhere in the Jura Mountains; not even in tunnels or in the core of deeply cut anticlines. No older formation was ever found than one special stratum of the Triassic. This special layer, the so-called anhydrite group, contains beds of rock-salt. Apparently, as suggested by Buxtorf, the anhydrite group played the rôle of a lubricant over which the overlying strata could glide and — detached from the underlying basement — were compressed into folds. Hence, these folds are considered as a surface phenomenon, a superficial type of folding which Swiss geologists have named a *décollement*. Due to the process of the Miocene *décollement* the super-structure suffered a compression of about 6 to 10 kilometres or 25 percent of the original breadth.

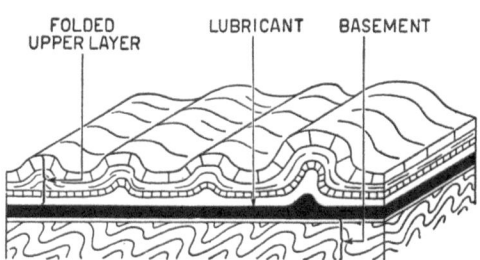

FOLDED     LUBRICANT   BASEMENT
UPPER LAYER

Fig. 45. "Décollement" of the Juras.

Fig. 45 shows this phenomenon in a schematic way.

Now that we know what the table is and how the tablecloth could glide over it the second question comes up: what element in the earth's crust pushed the tablecloth? In the opinion of Heim, Argand and many others the force came from the south. When the Alps were being compressed they are supposed to have overwhelmed the European block and to have pushed the Jura Mountains northward in front of them. Gliding over the lubricant layer the upper strata were compressed into folds over the unfolded basement consisting of Lower Trias, Permian, and crystalline rocks.

The arcuate chain of the Juras is framed by the upstanding basement blocks of Black Forest, Vosges, and Plateau Central. In between the two last named the high situation of the basement is revealed by a few smaller outcrops of crystalline rocks (fig. 42).

Apparently the site and outer boundary of the Jura Mountains are in some way related to the surrounding "framework" of elevated basement rocks, the movement of the surface layers having been hampered by and becoming adapted to the basement-blocks of the surroundings. In the depressed area of the Rhine graben the frontal chains of the Jura Mountains protrude farther northward than in the adjacent sectors which are opposite the higher blocks of Black Forest and Vosges.

On the other hand the whole arrangement of the Juras, as well as their time of origin, clearly shows a relation to the Alps. Two structural trends

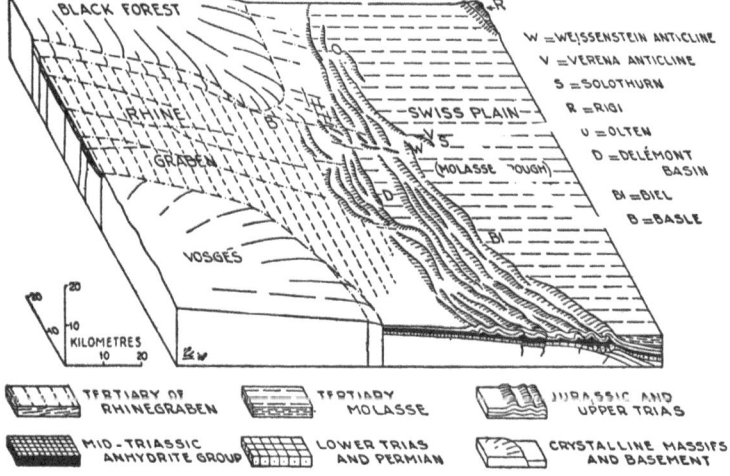

Fig. 46. Schematic block-diagram of the Jura Mountains between the Rhine graben and the Swiss plain.

occur in the Jura Mountains (fig. 46). In the region between Basle and the Swiss plain the dominant pattern consists of longitudinal folds with east-west trend; the second feature is formed by several short anticlines with a north-northeast to south-southwest trend. Thus the remarkable shape of the Delémont basin is due to a framework of folds consisting of these two elements. Another example is to be seen even as far south as the Verena anticline, a short and faulted anticline protruding from Molasse strata between Solothurn and the Weissenstein. The latter is one of the highest and southernmost longitudinal anticlines of this part of the Jura Mountains. Numerous faults run in the north-northeast direction. This direction is also displayed by the faults of the Rhine graben. Moreover they are of the same age, viz. they originated in Oligocene times, though some were rejuvenated in Miocene times. There even existed an open sea connection between the Rhine graben and the Molasse trough across the present area of the Jura Mountains in Oligocene (Stampian) times — the socalled Rauracian graben (fig. 42).

Apart from the northeast-southwest system some faults originated at right angles to this direction and these often include small rift-like blocks, which Glangeaud called *pincées*. The age of these is again Oligocene i.e. about 35 million years ago. On the other hand the east-west or longitudinal system of Jura folds originated towards the end of the Miocene i.e. about 15 million years ago. At that time the present pattern of two intersecting structural elements was completed. However, the transverse folds are quite different from the folds of the dominating longitudinal system. They are less numerous, they can be followed over a short distance only, and some of them show the characteristic arrangement which is called *en échelon*. Suppose two blocks separated by a fault and having their basement at different depths.

If the blocks moved southward, one over a greater distance than the adjacent block, a system of short *en échelon* folds would originate in strata overlying the fault.

Why did not the longitudinal system originate at the same time as the transverse system? We will proceed southward and see if an examination of the Swiss plain will furnish data which are relevant to the problem of the Juras.

### THE SWISS PLAIN

·The Weissenstein, one of the highest and southernmost Jura ranges, offers a magnificent view towards the mighty snow-capped chains of the Alps in the far distance. Did the Alps push up the Jura-chains? One wonders how this process took place, separated as the Alps are from the Weissenstein by a stretch of 20 miles of a low and rather flat country which is generally called the Swiss plain (Plate III, A). The Swiss plain is the surface of a trough which was gradually filled up with sediments as its foundation subsided. In the Lower Tertiary, approximately 35 million years ago, the Alps underwent one of their strongest phases of compression and folding. As soon as the strong compression decreased the folded mass rose up from deeper realms. At approximately the same time as the Alps began to rise, the area north of the Alps started its downward movement. However, the trough-shaped depression was constantly filled up by denudation products from the Alpine chains, which are called *Molasse* by Swiss geologists. The name Molasse means soft layers. The present Swiss plain is the surface of the filled up Molasse trough. The deposits

Plate III

(A) View across the Swiss plain towards the Alps, from the Weissenstein. The town in the foreground is Solothurn. See also figs. 42 and 46.

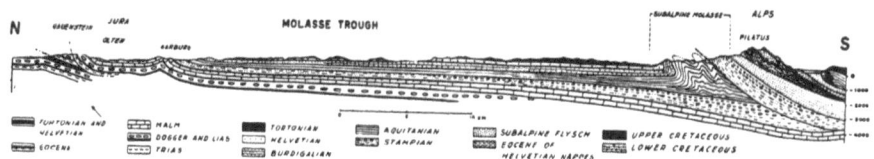

(B) Section across the Swiss plain. (After Baumberger).

consist of a sequence of Oligocene and Miocene sandstones and conglomerates, mostly rather friable (Plate III, B).

In the beginning the surface of the subsiding area had the aspect of a swampy plain. River deposits accumulated in large delta-like extensions in several places along the southern boundary of the lowland. At higher stratigraphic levels a few marine intercalations occur. In Miocene (Burdigalian) times, however, the area was largely invaded by the sea. Apparently the filling up with sediments was no longer able to keep pace with the rate of subsidence of the bottom at that time. Consequently marine Molasse follows on top of the lower freshwater Molasse.

The system of longitudinal Jura folds did not yet exist when the Molasse sediments accumulated. Their folding took place towards the end of the Miocene. As a matter of fact, Molasse sediments spread over large areas of the present Jura Mountains. Molasse deposits still occur in several synclines of the Juras and furnish a clue to approximately fixing the original northern boundary of the Molasse sea in Burdigalian and Helvetian times as indicated in fig. 42. With the exception of the area of the Rauracian graben their thickness is not appreciable as compared with the thickness of the deposits in the Molasse trough. Hence, the southern boundary of the Jura chains coincides approximately with the original boundary of the Molasse trough. A theory on the origin of the Juras ought to explain this coincidence. In the northernmost part of the sea some material was also transported from northern regions, especially from the Black Forest. However, even conglomeratic material of southern origin was deposited as far away from the Alps as the Jura region in Helvetian times.

Let us now examine the southern boundary of the Molasse trough. A schematic sketch of the distribution of strata along the surface of the trough shows the boundary region of Molasse trough and Alps to consist of a strip of Oligocene strata. Originally the southern boundary of the trough was farther south than it is now and also farther south than it was in Miocene times (cf. fig. 42). The present boundary is of tectonic origin due to "overthrusting" of the frontal chains of the Alps. In a few places this phenomenon can be seen very clearly, for example northwest of Champéry near Val d'Illiez where the Oligocene *molasse rouge* is found at a height of between 900 and 1500 metres underlying Pre-Alpine nappes.

Due to tectonic movements the deeper strata of the trough have been tilted and folded along the border of the Alps. Actually, without the occurrence of this disturbed subalpine strip our knowledge of the deeper layers of the trough would be much more limited.

Plate III, B, representing a section across the Molasse trough, according
to Baumberger, shows the disturbed southern part of the trough as well
as the frontal parts of the Alpine nappes resting on the Molasse.

This section is puzzling in more than one respect. One is the undisturbed
lower boundary of the trough as compared with the strongly disturbed surface features. The second question which comes up is this: was a push
from the Alps transferred through the Molasse sediments so as to cause the
folding of the Jura Mountains? Why then was only the southern part of
the trough influenced by this active force, whereas the sediments in the
northern, tapering part remained nearly uninfluenced or even quite
undisturbed?

So, the problem remains to be solved why the inner boundary of the
Juras coincides approximately with the outer boundary of the Molasse
trough.

According to the view already mentioned the tectonic disturbances
along the southern margin of the Molasse trough, too, are supposed to
have been caused by the pushing action of the Alpine nappes. As we shall
see, however, this theory is no longer tenable. In order to give a more
plausible explanation let us examine the structural history of the Alps.
Perhaps the Alps will also furnish a fresh outlook on the problem of the
folding of the Jura Mountains.

### THE HIGH CALCAREOUS ALPS AND THE "CENTRAL MASSIFS"

The surroundings of the Lake of Lucerne offer an excellent opportunity
for studying the contact between Molasse and the frontal ranges of the
High Calcareous Alps. Arbenz constructed a very lucid block-diagram of
the region (fig. 2, p. 4), the front section showing a profile along the
well-known Axenstrasse. Frontal lobes of two of the so-called Helvetian
nappes have been indicated by the numerals II and III.

They consist of Cretaceous strata, mostly limestones, which are surrounded by soft marls of Tertiary age, called *Flysch*. The northernmost element
of the Helvetian nappes is a slab of Cretaceous limestone resting against
a pile of tilted Molasse, though separated from it by a thin layer of Flysch.
The off-hand block represents the region of the famous Glarus-nappe (I),
the tectonic window of Elm, and a reconstruction of the Helvetian nappes
towards their roots behind the crystalline rocks of the Aar-Gothard
massifs (A). Finally, a few huge isolated blocks — the Mythen and several
others — may be seen. Their Mesozoic strata include facies types which
are entirely different from those found in the surrounding High Cal-

Plate IV

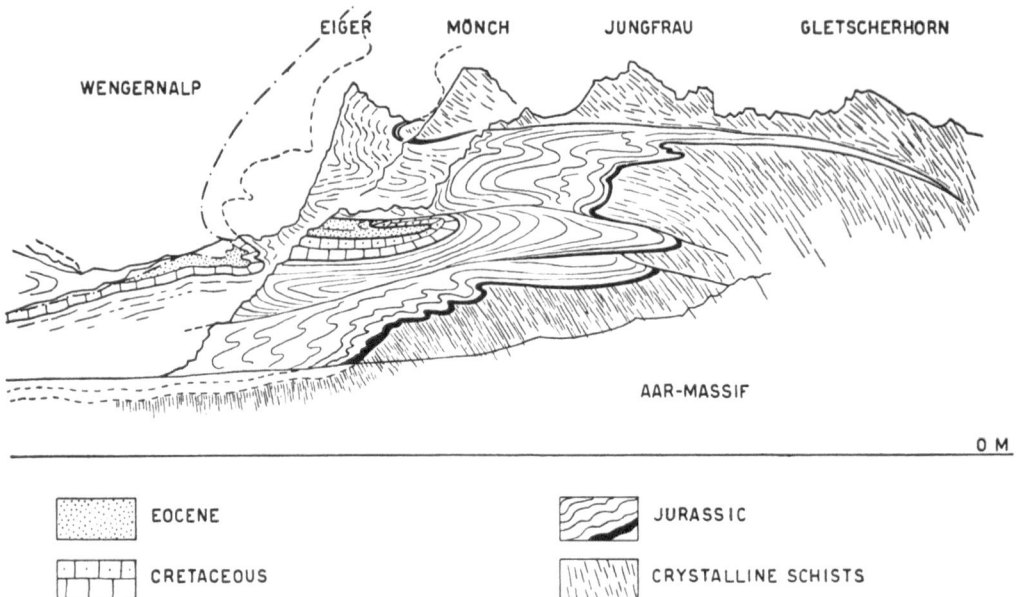

| | | | |
|---|---|---|---|
| EOCENE | | JURASSIC | |
| CRETACEOUS | | CRYSTALLINE SCHISTS | |

*(A)* Wedge-shaped structures in the Jungfrau massif (After A. Heim).

*(B)* Photograph of the same area, taken from Kleine Scheidegg.

careous Alps or in the Flysch on which they rest. Much larger units of the same strata occur in the Pre-Alps which will be considered further on. At any rate the strata of which these tectonic elements consist had their cradle still farther to the south than the Helvetian nappes.

Fig. 1 (A), p. 2, is a picture of the region represented in the north-western corner of the diagram. In the centre are the Mythen resting on a gently sloping landscape of Flysch. The frontal lobe of nappe III can be seen plunging below the level of the lake near Brunnen, and the northern-most slab of Cretaceous limestone is at the left-hand side of the scenery.

Now, according to a theory which for a long time was generally accepted this whole situation originated by a northward push from the Pennine nappes.

The Helvetian strata were squeezed out of their original sedimentation trough and thrust over the crystalline massifs (A in fig. 2, p. 4). At the same time the pre-Triassic crystalline rocks of the massifs were split up into wedges, some of which overthrust on Mesozoic strata. The Windgälle is a good example (see southern part of Arbenz' block-diagram, fig. 2). The same phenomenon can be very clearly made out from Wengern Alp when looking at the enormous high and steep northern front of the Jungfrau (Plate IV). Hence, there is no doubt about the great influence of pressure and push in this process. But recently Lugeon and Gagnebin pointed out that this was probably not the only factor concerned in the process of *mise en place* of the Helvetian nappes and the Pre-Alps. Gravitational downsliding of great slices, from the crystalline massifs towards the foreland of the Alps, is regarded by them as probably another factor that played an important role. One of their points is that some of the nappes are built up of soft material through which transfer of pressure over large distances cannot be assumed. Moreover, the frontal nappes consist of Cretaceous strata whereas the more southern nappes consist of Jurassic rocks. This remarkable fact can also be explained as a result of down sliding of the higher parts, leaving behind them the older strata which originally formed the core of the nappe.

In this connection the Glarus nappe has been a subject of much discussion. At the back on the right-hand side of his block-diagram (F in fig. 2). Arbenz indicated the tectonic window of Glarus schematically. Erosion of the Sernf-river cut through the Glarus nappe into the underlying Flysch, the nappe consisting of Permian conglomerates ("verrucano") resting on Jurassic limestone. For some time this structure was considered as the remnant of a large recumbent fold. Ampferer, however, regarded it as a result of downsliding.

It is difficult to give conclusive proof of the supposed downsliding process, though some examples in the Pre-Alps and the southern Alps (fig. 54 and 55) seem to be in favour of the downsliding theory, as we will see further on.

Probably, however, still another effect cooperated in the *mise en place* of the nappes. To render this clear we must return to the southern border zone of the Molasse trough. Once more we ask: Was it the pushing action of the Helvetian nappes that caused the Molasse sediments to become folded and tilted?

Was the heavy pile, some 2000 metres of conglomerates, of the Rigi forced up by the thin frontal sheet of limestone? (compare the left hand corner of the block-diagram fig. 2). It is hard to believe so, even if we realize that the erosional gaps were originally filled up by Flysch. And if it is true that the nappes came sliding down, the southward inclined position of their frontal lobes, including the limestone sheet, must be due to an additional and later tectonic process. Moreover, the Molasse conglomerates dip southward at an angle of 15° whereas the Flysch and Cretaceous limestone dip more steeply, up to 75°. It is obvious that the Molasse and Helvetian nappes are separated by faults, the existence of which was demonstrated by Buxtorf and his cooperators. Therefore, the conclusion seems inevitable that the disturbed position of the sub-alpine Molasse of the Rigi as well as of the frontal parts of the Helvetian nappes is due to tectonic action of their basement. The same holds good for the syncline of Amden, south of the Säntis mountains.

We need not regard the alleged tectonic action as due to a deep-seated and therefore unknown *deus ex machina*. For, in the nearby central massifs the basement of both the Helvetian nappes and of the Swiss plain is exposed. These massifs consist of a great number of wedges of crystalline rocks. The wedges were subjected to differential movements and the frontal parts of some of them reach as far northward as the Windgälle, Jungfrau, Breithorn and Muthorn, where they thrust over the Mesozoic and Tertiary rocks of the High Calcareous Alps. According to Günzler differential movements of these wedges played a role of importance in the northward movement of the Wildhorn nappe.

It would be absurd to suppose the wedges to be confined to those parts of the crystalline basement which happens to be exposed to view in the present state of erosion and denudation of the Alps. On the contrary, it seems hardly deniable that movements of similar wedges in the basement were responsible for the tectonics of the sub-alpine Molasse. Fig. 47, which is adapted from a series of sections by Günzler, illustrates this view in a schematic way.

The same fundamental idea was expressed by Bersier in a section across the Molasse trough near Geneva. In the southeastern part of Bersier's section the pushing action of wedges in the basement has disturbed the sub-alpine Molasse as indicated by arrows (fig. 48).

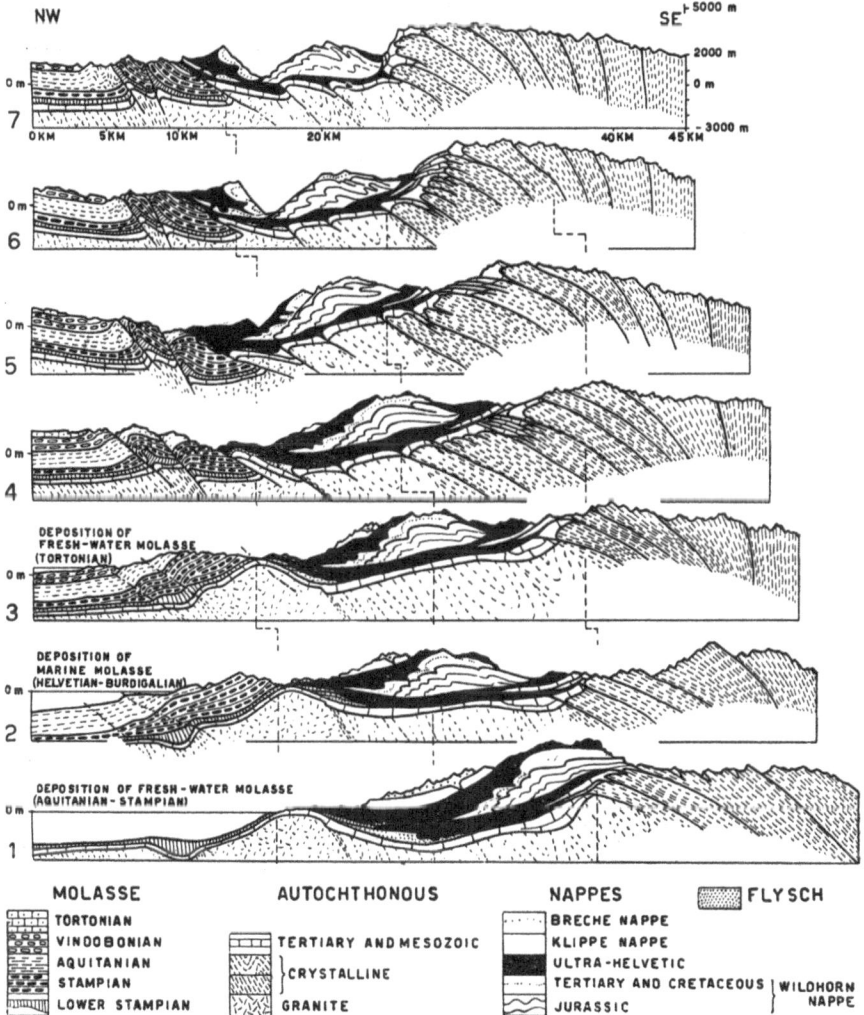

Fig. 47. *Mise en place* of the Wildhorn nappe and the action of crystalline wedges in disturbing the position of Molasse and Helvetian nappes (Adapted from Günzler).

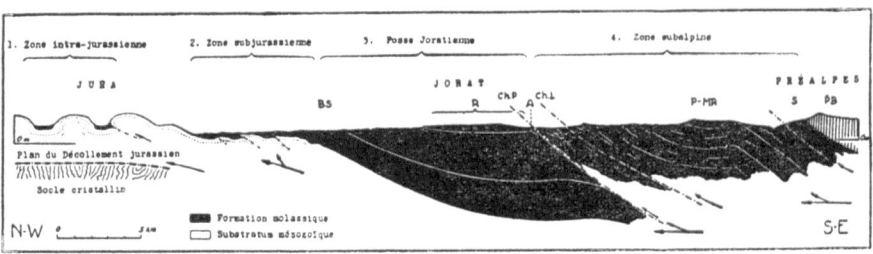

Fig. 48. Section across the western part of the Molasse trough near Geneva (After Bersier).

### AGAIN THE PROBLEM OF THE JURA MOUNTAINS

Bersier's theory is different from Heim's in that a tangential push from the south is supposed to be transmitted underneath the Molasse trough towards its northwestern border where it caused the folding of the Jura Mountains. With respect to this representation the question again arises: why did the Juras become folded while an intervening pile of Molasse strata remained nearly undisturbed?

How far the typical sequence of Triassic and Jurassic layers extends beyond the Jura Mountains to the S. and S.E. is unknown. The Trias of the Alps is developed in a quite different manner. However, to postulate that the lubricating layer coincides exactly with the present boundary of the Jura Mountains would be an entirely unfounded speculation. Besides it is not in agreement with known facts.

In the same manner the thickness and weight of the Molasse sediments cannot possibly furnish a full explanation because the Molasse trough extends much farther eastward than the easternmost Jura chains. Probably the processes involved were of a more complicated nature.

Lugeon went still further and considered the Jura Mountains as a nappe which slid into its present position under the influence of gravity. According to one suggestion a vertical uplift of the "central massifs" (Aar-Gothard, Aiguilles Rouges-M. Blanc) caused a downsliding of the plastic layers to the N. and N.W. The impulse is supposed to be transmitted below the Molasse trough. The masses sliding and accumulating under the influence of gravity are supposed to have caused the *décollement* of the Jura Mountains and their adaptation to the surrounding frame work of basement rocks. However, not only does the basement below the Jura Mountains dip towards the S and S.E., i.e. in the opposite direction to the supposed gliding of the Jura strata, but one might reasonably expect an enormous gap opened either between the Juras and the

Molasse trough or between the latter and the High Calcareous Alps as the 'tablecloth' slid away to the northwest.

Therefore Lugeon maintains that the pressing out of plastic layers beneath the weight of sediments of the Molasse trough in an outward direction must have caused the folding of the Juras. Even this suggestion does not explain the shortening of the upper structure as displayed by the Jura Mountains. Moreover, instead of folding in the Upper Miocene one might expect a gradual process keeping pace with the long continued accumulation of Molasse sediments during Oligocene and Miocene times and causing the squeezing out of the plastic layers along the outer boundary of the Molasse trough as the load increased.

In short, in the simile of the tablecloth being rumpled and forced over the table by some sort of pushing activity from the south, the following main problems remain unsolved: (1) what was the pushing element causing the *décollement*, (2) why does the inner boundary of the Jura Mountains coincide with the outer boundary of the Molasse trough, (3) why did the time of the *décollement* coincide with the Upper Miocene phase of diastrophism whereas such a phenemenon did not occur at the time of the much stronger Oligocene movements, (4) why did the greater part of the Molasse trough remain nearly undisturbed?

Theoretically there is a second possible way of rumpling a tablecloth. Imagine we keep the cloth fixed at a certain point and then pull the table southward instead of pushing its cover northward. The result would be the same. Though less easy to demonstrate the simile is probably more relevant to what actually occurred when the Jura Mountains originated.

Our present knowledge of the structure of the Alps leads to the conclusion that a process of major importance was progressive underthrusting of both the northern and southern "forelands" towards the central belt of the Alps. The same process probably played an important part in the formation of the Jura Mountains. If we accept a process of southward underthrusting of the basement, one question still remains unanswered, viz. why did the Jura strata not slip under the Molasse trough, or more precisely what was the obstacle that kept the "tablecloth" fixed at the present inner margin of the Jura Mountains?

The Oligocene movements in the Jura Mountains were contemporaneous with a phase of strong diastrophism in the Alps. However, only faults and some accompanying transverse folds came into being at that time. The dominating pattern of much stronger developed longitudinal folds originated during the more recent, though not so strong, phases of

Alpine compression in the Upper Miocene. There must, therefore, be a reason not only why the longitudinal pattern originated in the Upper Miocene but also why it did not come into being with the much stronger Oligocene phase of diastrophism. Apparently the factor — or factors — which caused the upper layers to become stripped off from their basement so as to result in the Miocene *décollement* was — or were — not yet present during the Oligocene diastrophism.

After the Oligocene diastrophism the Molasse trough originated. Therefore, it seems reasonable to suggest that this new element played a part. Moreover the Rhine graben began to form in the Oligocene, though probably a first downward movement had already started in the Eocene. However, the rise of the Rhine Shield which caused the formation of the Vosges and the Black Forest in the shape of elevated blocks took place in more recent times and had reached an appreciable amount in the Upper Miocene.

Hence, the influence of the updoming basement of the foreland is a second factor which needs further consideration.

During the formation of the Molasse trough the basement under the trough subsided by an amount of about 3000 metres. In this movement the marginal flexures were the weak strips predestined to become zones of tectonic disturbances during the subsequent epoch of diastrophism.

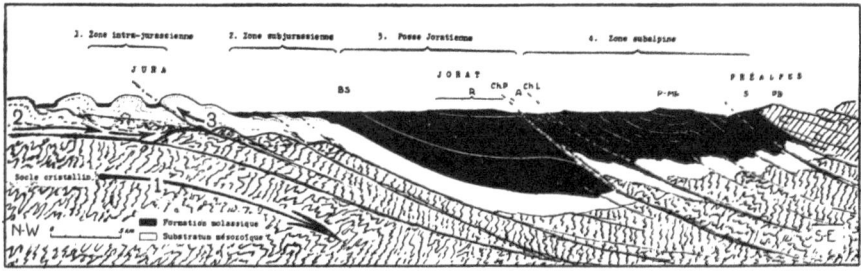

Fig. 49. Section across the western part of the Molasse trough; compare fig. 48.

Thus, the tectonic disturbances along the inner boundary of the trough may be considered as necessary consequences of the general southward underthrusting of the basement during the Upper Miocene phase. Proceeding northward the next zone of weakness is the outer margin of the Molasse trough. We may assume that the tangential force was transmitted mainly by the crystalline basement. The latter has been drawn only over a short distance in the left-hand part of Bersier's section fig. 48. If we complete it over the total length of the profile it will become clear that still another wedge had to be drawn, its front coinciding with the present southern boundary of the Jura Mountains (fig. 49). In doing so the first named problem is solved at the same time. For, it is now self-evident why the *décollement* started at this place and why the "fosse Joratienne" remained practically undisturbed. Let us accept, for a moment, one or more obstacles along the outer boundary of the trough which prevented the upper layers of the Juras from being dragged along when southward underthrusting of the basement occurred. Even then some difficulties remain, for the Molasse trough extends much farther eastward than the Jura chains. Therefore, the problem is only solved if a plausible reason can be given for supposing the basement wedges not to have originated along the outer margin of the trough eastward of the present Jura Mountains.

Perhaps the reason for the origin of basement wedges along the inner side of the present Jura Mountains as well as for their absence more eastward was the origin and extension of the other "new" element. Its influence has already been mentioned (p. 61). The shape of the Jura arc is adapted to the surrounding outcrops of basement rocks. Its eastward end corresponds exactly to the eastern boundary of the Black Forest (fig. 42), and the northernmost Jura folds protrude farther northward in the Rhine graben as compared with the adjacent sectors opposite the Vosges and Black Forest. Hence, it seems clear that without the presence of the highly upraised basement rocks the Jura arc would not have formed.

It must be realized that only in some special areas have the basement rocks been elevated so high as to form large horst-like blocks like the Vosges and the Black Forest. However, a general tilting movement took place in a much wider surrounding area. Even in the Rhine graben the renewed movements along faults that occurred in Aquitanian times were accompanied by elevation. In this manner the surface of the basement under the Jura strata gradually came to slope upward in an outward direction. This holds good also for the western part of the Juras. For, even in this region, where the Juras are separated from the Plateau Central by the Saône-Rhône depression, outcrops of basement rocks are found along the outer margin of the Jura arc (fig. 42). Though the outer boundary of the Juras has the general appearance of a gently curved arc, it actually consists of two arcs (fig. 42). According to Lugeon a southwestern arc called *arc Lédonien* reaches as far as Salins whence a second arc, the so-called *arc Bisontin*, can be traced eastwards. Possibly this phenomenon is related to the differing amounts of tilt of the basement under the two respective arcs, the surface under the Lédonian arc dipping at a slightly greater angle than that of the basement under the "arc Bisontin".

Apparently subsidence of the Molasse trough was not sufficient to predestine the origin of basement wedges along its outer boundary. Only

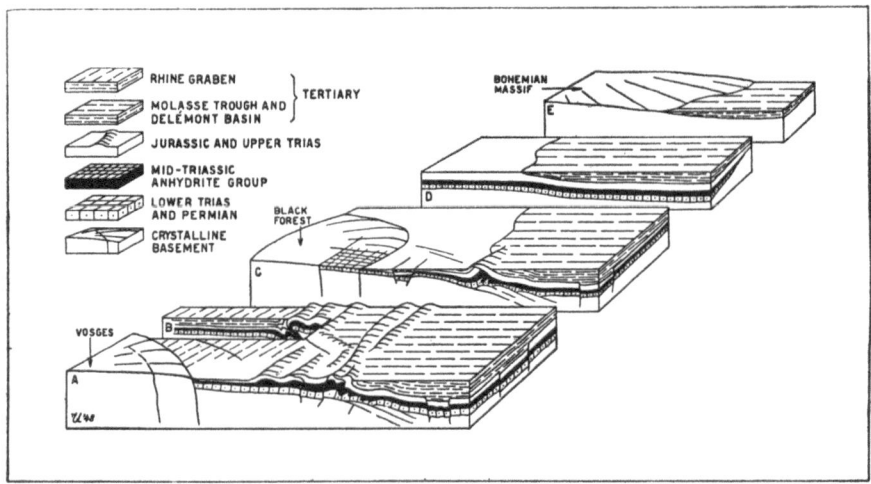

Fig. 50. Schematic representation of different sectors between the Vosges and the Bohemian massif.

when, at a later time, the outer boundary of the trough became the inner boundary of an area with opposite movement — i.e. of an upward tilting of the present area of the Juras — the hinge line between the two opposed movements was weakened to such a degree that basement wedges and consequently a *décollement* could originate. Tentatively the situation is represented by blocks A–D of fig. 50. The tilted basement is shown by blocks A, B and C.

Block D represents a schematic profile east of the Black Forest. There the basement has not been tilted in an outward direction. Consequently no basement wedges originated and no *décollement* took place.

Still further eastward again there is an elevated block of basement rocks, the Bohemian massif (fig. 50, block E). There, however, neither Jura strata nor lubricating layers are present, and the prism of Molasse sediments is thinner and rests directly on the granites and gneisses of the Bohemian massif.

Moreover, it is the upward slope of the basement that — combined with four other factors — was of deciding importance in controlling the outer boundary and shape of the Jura arc as well as the character of the folding.

The four other factors were: (1) the lubricating layer, (2) a certain thickness of the upper structure, (3) the limited space in transverse direction due to the obstacles along the outer boundary of the Molasse trough, and (4) Alpward underthrusting of the basement.

In the Rhine graben the basement is at a lower level than on the adjacent blocks. It is for this reason that *ceteris paribus* the outermost folds of the longitudinal system occur farther northward than on the higher blocks on either side where similar conditions controlling the outer boundary of the arc are to be found more to the south (see fig. 50 blocks A, B, C).

In short, the fact that a structure like the Jura Mountains is an exceptional feature means that its formation is due to the accidental presence and interaction of several factors. If we accept a southward movement of the basement the evident reason why the layers above the lubricant layer became folded within the limited space of the Juras, is the presence of the upward slope of the basement rocks in an outward direction. In this process however, certain other factors were responsible for the coincidence of the inner margin of the Juras with the outer margin of part of the Molasse trough.

Generally, the wedge-shaped elements have been considered as a result of northward overthrusting. More probably, however, the main factor responsible for their origin was a process of progressive underthrusting of the foreland towards the Alps.

Eventually, at the end of our route across the whole complex of Alpine chains, we will find convincing evidence for this phenomenon of underthrusting of the basement towards the Alps.

### THE PENNINE ALPS

After having crossed the High Calcareous Alps and the Central Massifs we reach the very heart of Alpine tectonics. The frontal parts of the enormous Pennine nappes were hampered in their development by the huge blocks of the Aar and Gothard massifs. More to the west, the surface of these massifs plunge towards greater depths to reappear in similar masses like Montblanc-Aiguilles Rouges and Pelvoux-Belledonne. The deeper these wedges are situated the more chance had the Pennine nappes to move northward. Looking at a structural map (fig. 42, p. 56) one will notice an outward bulge of the Pennine nappes between the Aar-Gothard and Mont Blanc-Aiguilles Rouges massifs. In this area the front of the Pennine nappes was pressed against the Helvetian nappes. The Rhône-valley follows this boundary zone between Brig and Saxon. The northern flanks of the valley, composed as they are of limestones and marls of the Wildhorn and Diablerets nappes, strongly contrast with the southern flanks of Pennine gneisses and their enveloping *schistes lustrés*. The boundary runs right across the town of Sion.

The Pennine Alps belong to one of the classical terrains of structural geology. To illustrate this statement let us mention only two examples.

It was in this area that the driving of the Simplon tunnel brought conclusive evidence in favour of the structural conceptions put forward by Schardt and Lugeon. The Simplon region between the Rhône valley and the Italian Val Divoria is made up of three huge recumbent folds — named

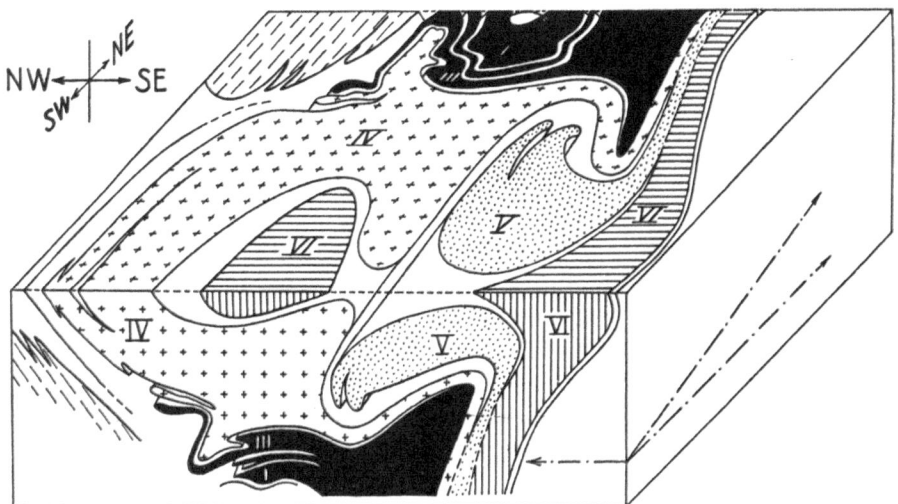

Fig. 51. To illustrate Argand's principle of axial elevation of the Pennine nappes.

Antigorio, Lebendun, and Monte Leone — consisting of pre-Triassic gneiss-cores surrounded by and enveloped in their covering strata of Triassic and Mesozoic *schistes lustrés*. More to the west, in the surroundings of Zermatt, Argand tried to unravel the intricate tectonics of three other Pennine nappes — called St. Bernard, Monte Rosa, and Dent Blanche — and in a brilliant synthesis he suggested the three Simplon nappes pitching laterally below the higher nappes of Zermatt. He was guided by the so-called principle of axial elevation and depression of the nappes behind the Aar-Gothard massifs. In this manner he was able to construct a section of the Zermatt region showing the tectonic elements down to a depth of about 20 kilometres below the present surface (fig. 51).

Everybody who has visited the Pennine Alps and experienced the difficulties inherent in the intricate stratigraphic and structural relations of the region as well as in its rugged though magnificent scenery will always retain deep respect for the achievements of the Swiss geologists who succeeded in unravelling the tectonic structure.

The structures revealed in photographs of two views in the neighbourhood of Zermatt find their geological explanation in the accompanying geological section of Plate V. In the Austrian Alps still higher tectonic units were found. Their major elements have been called Err-Bernina, Campo, and Silvretta-Oetztal nappes. According to general opinion the Tauern mountains in Austria represent a tectonic window in which Pennine nappes have been laid bare by erosion whereas elsewhere in Austria the Pennines are still covered by the so-called East Alpine nappes. According to some authors these nappes originally covered also the Swiss region. Fig. 53 is a synthetic picture which incorporates this opinion. Eventually we shall return to this question. Table II is a greatly simplified synopsis of these ten Pennine and East Alpine nappes. They are usually indicated by the Roman figures I–X. The same indications are used in the tectonogram Plate VII.

Table II. *Pennine nappes of Switzerland and higher tectonic units.*

| | | | | | |
|---|---|---|---|---|---|
| IX. Silvretta, X Oetztal | Upper | | East Alpine nappes | | |
| VIII. Campo | Mid | | | | |
| VII. Err Bernina | Lower | | | | |
| VI. Dent Blanche | | | | | |
| V. Monte Rosa | Michabel | | | Suretta | in |
| IV. St. Bernard | nappe | Pennine | | Tambo | eastern |
| III. Monte Leone | | nappes | | Adula | Switzer- |
| II. Lebendun | Simplon | | | | land |
| I. Antigorio | nappes | | | etc. | |

It remains to be seen how far the different nappes extend laterally. Moreover one would not be surprised if the shape of the nappes and folds changes considerably within comparatively narrow limits due to both variation of their original size and local disturbing influences. The same uncertainties are inherent to the principle used by Argand in the construction of his deep reaching profile (fig. 51). It ought to be considered *cum grano salis* inasmuch as it incorporates the doubtful assumption of uniformity of thickness and shape of the nappes over large distances.

The same principle was carried *ad absurdum* by Staub in longitudinal

sections through the Alps showing — for example — the Simplon and
Monte Rosa nappes like undulating blankets over a distance of more than
900 kilometres!

In 1916 Argand tried to trace the structural history of the Swiss Alps
from Carboniferous times onwards. One of the principal ideas of this
representation (fig. 52) is the subdivision of the original geosyncline by
the development of submarine ridges and intervening secondary sedimentation troughs. Out of these embryonic ridges some of the nappes are

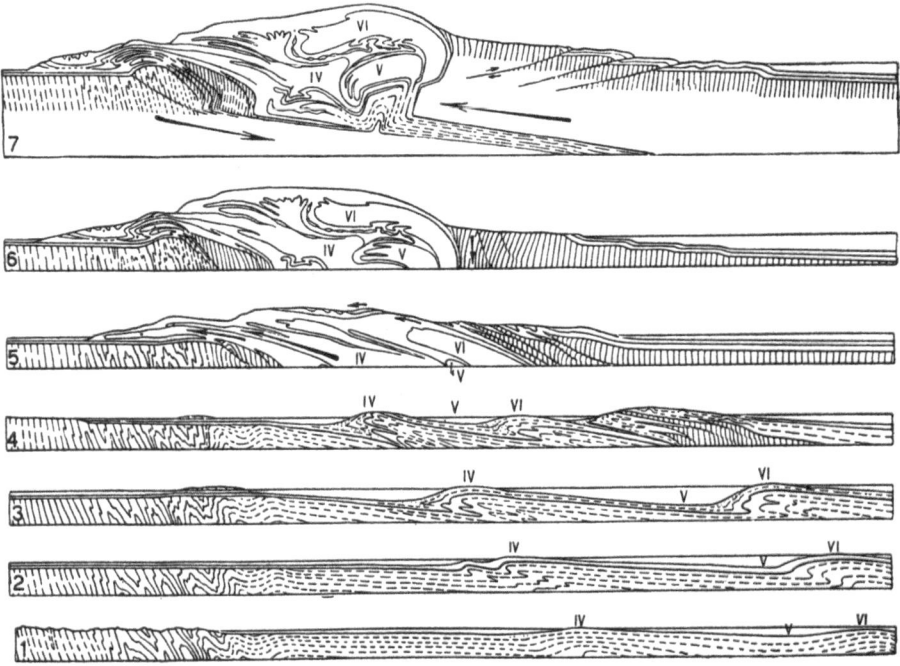

Fig. 52. Origin and development of the Swiss Alps (After E. Argand, simplified).

thought to have developed, like the St. Bernard and Dent Blanche nappes,
whereas the Monte Rosa nappe rose up out of the bottom of the depression
in between them. The subsequent profiles in this motion-picture show
the gneiss core of the Monte Rosa nappe being pushed upwards when the
two adjacent geanticlines had already travelled a long distance towards the
northwest and had attained great dimensions.

At this stage the Monte Rosa core advanced. By lack of space it was
forced to intrude into the rear parts of the St. Bernard nappe causing the
development of the Michabel back fold (profile 7, fig. 52). The backward
folding of the St. Bernard nappe, at any rate the upper part of this same pheno-

Plate V

(A) View from the Gornergrat, near Zermatt. Dotted lines indicate boundary of Monte Rosa nappe (Monte Rosa to the left) and overthrusting of Dent Blanche nappe (Matterhorn).

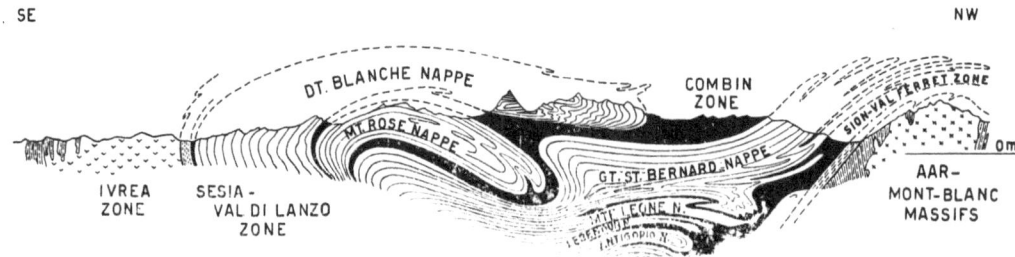

(B) Generalized section across the Pennine nappes near Zermatt. (After E. Argand).

(C) The famous backfolding of the Michabel (Great St. Bernard nappe) northwest of Zermatt (the village in the foreground). DB, Dent Blanche thrustplane; St. B., St. Bernard thrustplane.

menon, can be seen in the mountains at either side of the north-south running valley of Zermatt. The west side is shown in Plate V, C (the Michabel group is at the opposite side of the valley).

The gneiss core of the St. Bernard nappe is separated from the overlying Dent Blanche nappe by *schistes lustrés*. Indeed, this is the case in the valley of Zermatt and Argand thought the same relationships would persist still much farther eastward along the whole northern and eastern outcrops of the Monte Rosa core.

This expectation does not appear to have realised. In 1935 Huang, finding the *schistes lustrés* to be lacking over some distance in the boundary region between the St. Bernard and Monte Rosa cores thought this might be explained by these rocks having been squeezed out. Afterwards Bearth mapped the critical region again and examined the rocks petrographically in great detail. His conclusion is that there does not exist a tectonic boundary and no petrographic differences whatever between the St. Bernard and Monte Rosa nappes over a distance of $3\frac{1}{2}$ kilometres. Hence ,they have to be united in one larger structure which is called the Michabel nappe. This interpretation is now accepted on the official geological map of Switzerland. Bearth's results have been worked up in our schematic tectonogram Plate VII. Though Bearth's tectonic map is different from Argand's only in a few details, it affects fundamental parts of Argand's structural and genetic concept of the Pennine nappes.

In the first place the new data are fatal to the interpretation of the famous *pli en retour du Michabel* as put forward by Argand. Moreover, it is difficult to reconcile the new section across the Zermatt region with Argand's series of pictures (fig. 52) in which he condensed his ideas about the birth and evolution of the Pennine nappes.

The reconstruction of the original ridges and intervening troughs (fig. 52, section 4) was based mainly on the distribution of breccias in the Trias. Their occurrence was thought to correspond with areas which originally were near a ridge protruding above sea-level. Their absence suggested deposition in an area between two ridges. Was the meaning of these data overrated by Argand? Lacking sufficient data this question cannot be answered and we must wait still further detailed investigations in the interesting area of the Pennine Alps. The structural history appears to be still more intricate than could have been foreseen a few decades ago. To be sure, one need not be surprised if further explorations will involve still more radical changes of Argand's structural picture of the Pennine Alps.

### THE PRE-ALPS AND THEIR "ROOTS"

Another prominent feature in Argand's synthesis is the suggested northward overthrusting of the southern "hinterland" of the old Tethys geosyncline over its contents and its northern foreland. In its most extreme expression it was formulated as overthrusting of Africa over Europe. As a matter of fact the crystalline cores of the Austro-Alpine nappes were regarded as frontal parts of the hinterland. This idea has been generally accepted by Swiss geologists. It is the leading theme in Staub's *Der Bau der Alpen* and it was also propagated by Collet in his well known book *The Structure of the Alps*. Thus, for example, Collet writes: "the higher Pre-Alps, that can be seen from Geneva, Lausanne and Berne, represent a small part of Africa resting on Europe or Eurasia". (See fig. 53).

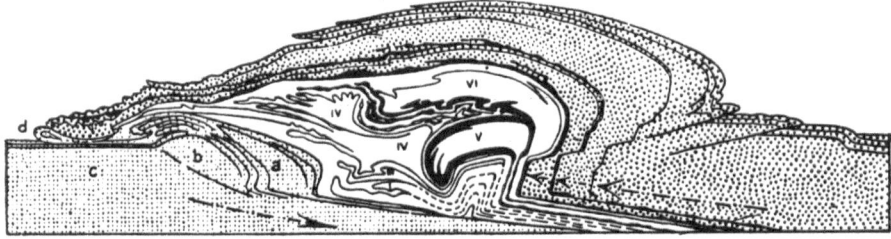

Fig. 53. Generalized section across the Alps. The foreland Europe (c) includng the crystalline wedges (a) and (b), is indicated by small dots. Larger dots: the "hinterland" Africa including the large overthrust slices of the East-Alpine nappes, and the Pre-Alps (d). I–VI are the Pennine nappes. Basic rocks are in plain black notation. (After E. Argand).

However, recent investigations clearly show this idea to be untenable.

In the first place equivalents of the "Préalpes médianes" are known from French territory where they partly correspond with the "klippes de Savoie". French geologists, however, always had good reasons for quite different views on the structural development. In their opinion these masses originated from troughs along the external or convex side of the Pennines (zone subbriançonnaise). Opinions changed so to say at the Swiss frontier mainly under the influence of the tectonic interpretation of Argand and Staub. But, recently, Tercier was one of the first Swiss geologists to dissent. On account of detailed stratigraphic investigations in the *Préalpes médianes* he has given strong arguments in favour of the rival theory. The sediments of the Pre-Alps, he maintains, originally accumulated in the northern part of the Pennine region.

Crystalline schists of the Ivrea zone (fig. 53, cf. fig. 42) considered by Argand as the roots of these Pre-Alps, were called lower East-Alpine nappes by Staub. A quite different interpretation of the Ivrea zone was given by other workers who considered this zone of schists as a mass of pre-Triassic rocks more or less comparable to the Aar and Gothard massifs from a structural point of view. The northern boundary of the Ivrea zone is characterized as a strongly mylonitized zone of tectonic movements.

Apparently the Ivrea-Insubric zone is not a zone of roots of nappes and E. Haug was right when as early as 1925, he attempted to demonstrate that Switzerland was never covered by East-Alpine nappes. Results obtained in the Bergamasc Alps have given conclusive evidence, as we shall see presently. The tectonic crush-zone along the northern boundary of the Ivrea and Iorio strip can be followed further eastward where it is generally called the Insubric line. The East Alpine nappes occur north of the Insubric line whereas the Lombardic Alps are to be found south of the same line. Another important tectonic line which will come up for dis-

cussion forms the eastern bounda-ry of the Lombardic Alps: the so-called Judicaria fault line (fig. 56).

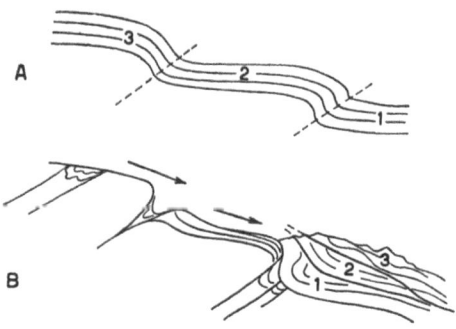

Fig. 54. Schematic representation of the principal movements in the southern Alps showing un-derthrusting of the crystalline basement in the Lombardic Alps, and downsliding of their sedi-mentary cover.

The key to the tectonics of the Southern Alps may be condensed into a simple formula, which is represented in fig. 54. The crystal-line basement forms a few huge stepping-stones leading from the plain of Lombardy in the di-rection of the Insubric line. Mostly the flexures became too steep and gave rise to wedge shaped structures. This is the interpretation given by de Sitter in a paper summarizing in a clear way the work of the Leyden School, achieved in the Lombardic Alps during the last decades. According to Lugeon, Gagnebin and others downsliding of nappes over rather great distances must have been a frequent process in the Helvetian Alps and the Pre-Alps (see p. 65). If we look for similar features elsewhere we might expect to find them in the Lombardic Alps. Now, indeed, it seems to me hardly possible to show a more convincing example of gravitation tectonics (downsliding nappes) than the nappes shown in fig. 55.

De Sitter's geological sections agree in essentials with a profile across the Lombardic Alps published by Dozy in 1935 (fig. 55). This author already argued that the crystalline rocks in the northern part of the Lombardic Alps cannot possibly be regarded as roots of East Alpine nappes as supposed by Staub. On the contrary, his profiles show the Orobic zone to be influenced by numerous southward overthrusts.

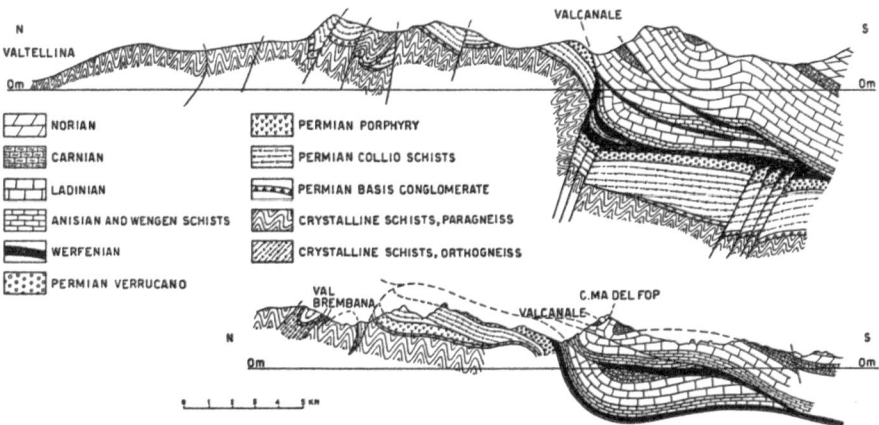

Fig. 55. Two geological sections across the Bergamask Alps (Lower after de Sitter, Upper after J. J. Dozy).

In 1947 de Sitter gave a reconstruction of the troughs in which the Mesozoic sediments accumulated. Three depressions were separated by two intervening ridges viz. the Averara ridge and the Camonica ridge (fig. 56). Moreover he made it clear that these ridges and troughs had a southwest-northeast trend, though the tectonic structures of the Lombardic Alps are rigorously east-west.

Still another feature of great importance was pointed out by the same author. By a comparative analysis he found three stratigraphic units of the Southern Alps to have their counterpart in sediments of three major units of the East Alpine thrust sheets (fig. 56) viz. (1) Err-Bernina (2) Campo and (3) Silvretta respectively. Therefore, he suggested that the Mesozoic sediments of the East Alpine nappes originated in a series of troughs which were the continuation of the Lombardic troughs. His schematic synthesis is shown in fig. 56. Apparently the original trend of the troughs and intervening ridges from which the Lombardic Alps were born is intersected at a sharp angle by the Insubric line. During the Alpine paroxysm the big Insubric fault-zone originated, cutting obliquely across the existing troughs and ridges. Their northeastern portions overthrust towards the north and became the East Alpine nappes, whereas the

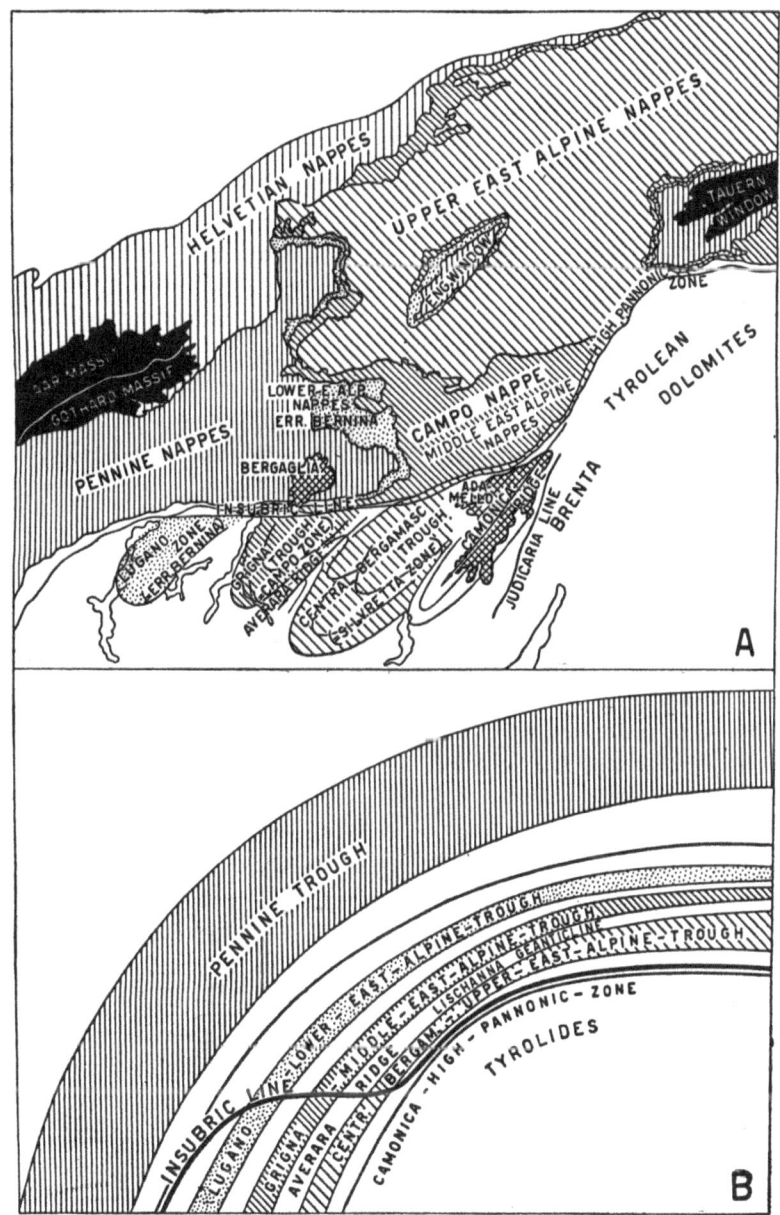

Fig. 56. Tentative and schematic construction of the sedimentation troughs of the Alps.
(After de Sitter).

southwestern part formed the great blocks and wedges of the Orobic zone and the Lombardic Alps.

The Judicaria line is considered to be a transcurrent fault of large dimensions. So much seems established without doubt: the region south of the Insubric line did not give origin to overthrust sheets towards the north or northwest. Switzerland is not a "half window" in the sense of Staub. A *fortiori* there is no question of overthrusting of a hinterland ("Africa") over the Alps. Whatever the interpretation of the East Alpine nappes, they originated from the Tethys geosyncline itself which apparently consisted of an intricate pattern of troughs and intervening ridges.

From a structural point of view the region south of the Insubric line is comparable to the northern Hercynian massifs like Aar-Gothard and Aiguilles Rouges-Mont Blanc. Both are blocks of the pre-Triassicbasement split up into numerous wedges showing differential movement and dipping mainly towards the original geosyncline.

### PLUTONIC BODIES

We have crossed Switzerland from the Juras to the Southern Alps. Before ending our trip, however, we still have to make an excursion sideways towards the granite massifs of Bergell and Adamello. Those who continue with us to the end will notice that the route leads to problems of the deep realms below the Alps.

Uniting the results arrived at so far, a schematic and generalized section across the Alps ought to express underthrusting towards the mountain-chain from both the northern and southern foreland. Other considerations strengthen this conclusion. Unrolling of the Alpine nappes reveals a transverse shortening of the whole chain by at least 200 kilometres.

Evidently, the crystalline basement must have undergone shortening by the same amount. It will be argued --- in chapter V --- that the total mass of basement rocks represented by crystalline massifs and the crystalline cores of nappes is far from sufficient to explain a crustal shortening of 200 kilometres. Hence, the basement must have slid downward and been doubled up under the Alps so as to form a sialic mountain root. The problem of the dimension and shape of the mountain root is a topic that will be taken up in Chapter V. At present, however, we have to consider a process which is intimately related to the forcing down of a root of comparatively light, so-called sialic rocks under the folded contents of the geosyncline. Due to thermal processes the root probably became mobile and part of the material migrated upwards. According to several authors (cf. p. 139) it is to the uprising material that we must ascribe the widespread metamorphism of the folded rocks of the geosyncline. It may lead to migmatization and ultimately to the emplacement of granite bodies, so-called granite batholiths (cf. p. 139). The situation of the granite and tonalite intrusions in the Ivrea-Insubric zone and its northern vicinity is shown on Plate VII.

The main occurrences in the area under discussion are Biella and Traversella, Bergell and Adamello, the latter is situated where Insubric line and Judicaria fault meet (fig. 56). As shown on the block-diagram, Plate VII, in a schematic way the Bergell granites have pierced through the Alpine nappes. Evidently, the process must have taken place after the nappes had reached their present positions. Consequently, the granite is of Tertiary age. Some authors even think the process of acid intrusion ended during the rising movement of the Alps in the Upper Tertiary or even later, though the process may have started much earlier in deeper realms and gradually worked its way upwards.

The situation of these young intrusive masses along the Insubric zone or in its vicinity as well as their absence along the boundary of the belt of northern or so-called Central Massifs is quite understandable inasmuch as the southern zone of crystalline wedges is steeper than that in the Central Massifs and so facilitated the upward migration of material.

Much older than the acid intrusions are the basic rocks (cf. p. 138) which are widely spread in the Pennine Alps, where they form numerous intercalations in the *schistes lustrés*. According to recent data intrusion of the basic rocks, now mostly called ophiolites (spilite, gabbro, basalt and serpentine) took place in Jurassic and Cretaceous and up to Eocene times.

So, the two main groups of "igneous" rocks differ also in the place and the time of their formation. Probably the basic rocks originated as abyssoliths when basic to ultrabasic magma got an opportunity to ascend from deep realms along deep-reaching faults and thrust-planes. On the other hand, the origin of the Tertiary granites and tonalites was possibly due to processes starting from the sialic mountain root (see fig. 9 and cf. p. 138–140).

Apart from the young granites considered so far there are several other granites in the Alps, but these belong to the Variscian or older orogenic cycles. This is undoubtedly the case as regards the granites of the Central Massifs. The granite of Mont Blanc, for example, is post-Carboniferous and pre-Triassic.

Probably also the granites of some of the East Alpine nappes are very old.

## BACK TO THE JURA MOUNTAINS

Our excursion to the plutonic bodies of Bergell and Adamello brought us to deep realms under the Alps. One of the conclusions was that a mass of comparatively light crustal material must have been forced downwards under the Alps due to a process of progressive underthrusting from both the northern and the southern forelands. At last, after having crossed the Alps entirely, we found the evidence which we wanted so eagerly when discussing the origin of the Juras (p. 73).

The *décollement* of the Juras may be described in various ways. One extreme is to suppose a movement of the upper layers towards the foreland over a stable basement. The contrasted surmise is the suggestion of an Alpward movement of the basement combined with an obstacle which prevented the upper layers from partaking in it beyond a certain limit. Making a choice between these two possibilities would be an arbitrary procedure if the Jura Mountains were considered separately. The second point of view is, however, found to be the more probable one when the Juras are considered in connection with the evolution of the deeper structure of the Alps as a whole. For, the probability of progressive underthrusting from both the foreland and the hinterland of the Alps towards the central zone of the mountain-root was based on converging evidence from geological sources.

Additional evidence from geophysical data will be given in Chapter V.

Therefore, the moment has come now to return to our starting point, the Jura Mountains, in an attempt to picture their origin in connection with the evolution of the Alps.

The suggested sequence of events that gave rise to the Jura Mountains is represented schematically in fig. 57 whereas the tectonogram, Plate VII, shows the area under discussion in its relation to the structural elements of the Alps.

It is from the Eocene that the first signs of the formation of the Rhine rift valley are known. Very strong compression moulding the Pennine nappes into their present shape had finished before Rupelian times.

Subsidence and movements of faults in the Rhine graben were accentuated during the Oligocene. Differential movements along blocks caused the roughly N–S pattern of short anticlines in the area of the present Jura Mountains.

Block I of fig. 57 represents the situation after the Oligocene diastrophism. Following on this paroxysm both the present Lombardic and Molasse troughs started their downward movement in the Oligocene (Rupelian).

Probably their subsidence kept pace with the rising movement of the folded belt in between them. A causal relation between the two phenomena seems highly probable.

The so-called Rauracian graben is regarded as a southern continuation

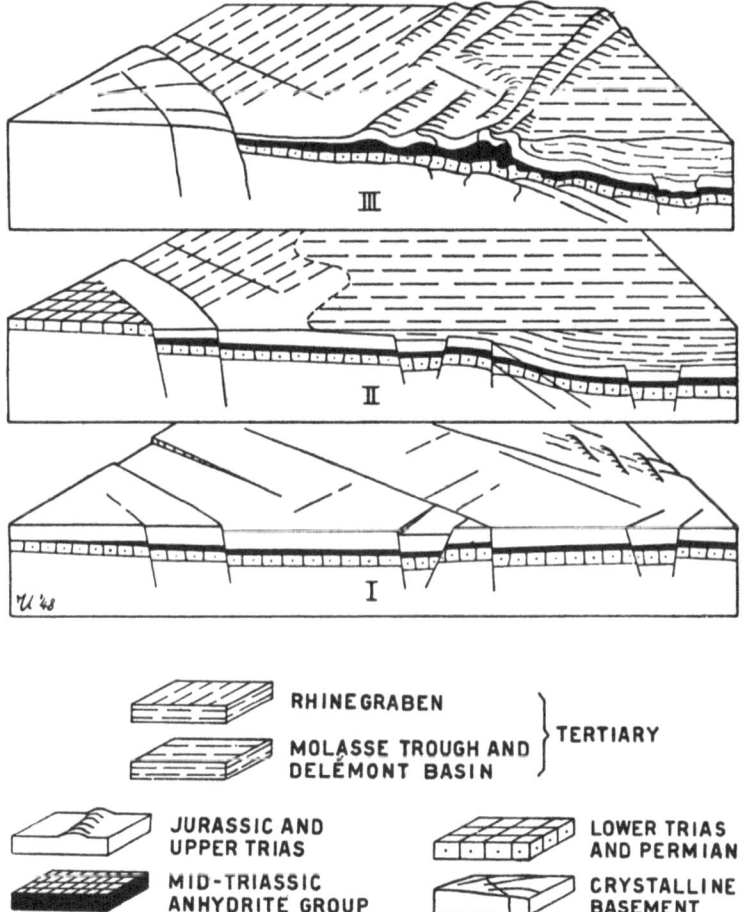

RHINE GRABEN
MOLASSE TROUGH AND DELÉMONT BASIN } TERTIARY

JURASSIC AND UPPER TRIAS
MID-TRIASSIC ANHYDRITE GROUP

LOWER TRIAS AND PERMIAN
CRYSTALLINE BASEMENT

Fig. 57. The suggested sequence of events that gave rise to the Jura Mountains.

of the Rhine rift valley connecting it with the Molasse trough. The basement became gradually tilted-up outward and in part elevated to form the high blocks of the foreland (Black Forest, Vosges, Plateau Central) as well as the smaller outcrops in between them. In the meantime denudation products from the Alpine chains filled up the subsiding

northern and southern troughs and spread far northward into the area
of the present Jura Mountains (fig. 57, block II). Renewed compression
towards the end of the Miocene involving underthrusting of the basement
towards the Alps, had again several far reaching consequences. The central
and southern massifs became more accentuated. In consequence, Tortonian
strata of the Lombardic trough were overrun by Triassic strata. As
another consequence the Helvetian nappes came to overrun the Molasse
deposits. Gravitational downsliding possibly played a role of some
importance in this process. Differential movements along planes separating
wedge-shaped parts of the basement caused the southern boundary zone
of the Molasse trough (including the frontal part of the Helvetian nappes)
to become tilted and adjusted into their present position. Due to the
same general process of underthrusting another set of faults and intervening
basement wedges originated along the next zone of weakness, viz. along
the outer boundary of the trough (fig. 57, block III).

However, they originated only where the flexure became weakened by the
subsequent upward tilting of the basement in the adjacent surrounding area.
Without this additional weakening of the flexure no basement wedges
and therefore no Jura Mountains would have originated (cf. fig. 50 D).

The wedges formed an obstacle which prevented the strata above the
lubricating layer of the Mid-Triassic from being dragged along southward
and downward so as to partake in the movement of the basement underthrusting towards the Alps. Possibly the prism of Molasse sediments
formed an additional factor. The superstructure gliding over the lubricating layer became stripped off and rumpled into folds which in their
general arrangement had to adapt themselves to the elevated basement.
Possibly the connection between the lubricating layer beneath the Juras
and its continuation under the Molasse trough was already broken or
squeezed off during the subsidence of the trough and the additional
tilting of the surrounding foreland. If so, this was a third reason for the
coincidence of the outer margin of the Molasse trough and the inner
boundary of the *décollement*. The wedges are drawn very schematically
in the block-diagram. Undoubtedly the situation is much more complicated,
and possibly it would be better to imagine zones of imbrication instead
of wedges which are a too simplified representation.

It seems improbable that the basement under the Juras is so smooth
as was originally supposed. On the contrary, structural elements of the
basement dating from Oligocene times (some even from earlier times)
presumably had a great influence in the development and arrangement of
the Upper Miocene folds. Several transverse folds reveal the influence

of an older pattern of the substratum in a convincing manner. According to Liniger the anticlines which surround the Delémont basin existed even in pre-Stampian times, but they became rejuvenated subsequently, once in the Oligocene, and once again in the Miocene. Some of the Oligocene transverse faults became rejuvenated during the Miocene phases of movement. Among them is the well known set of transcurrent faults. Probably, however, several longitudinal elements of the basement also had a great influence in the arrangement of the Upper Miocene pattern of folds.

The possible influence of basement wedges was suggested by Aubert whose opinion is clearly represented in fig. 58. If such structures exist, their origin is probably due to the same phenomenon of Alpward under-

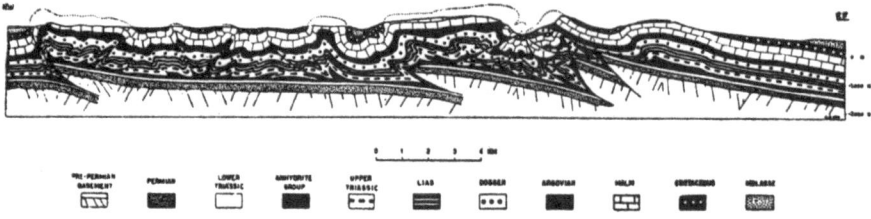

Fig. 58. Wedge-shaped structures in the basement of the Jura Mountains. (After Aubert).

thrusting of the basement that caused similar though larger wedges along the inner margin of the Juras. If so, they must also have originated in the Upper Miocene. Another possibility is that longitudinal faults and relatively small graben-like structures (the *pincées* of Glangeaud) dating from Oligocene or even older times had a great influence in the arrangement of the Upper Miocene pattern of the Juras. This idea is substantiated in the schematic and tentative block-diagrams of fig. 50 and 57. Most probably both features played a rôle of importance.

We have found, after all, that the folding of the Jura Mountains appears to have been a far less simple process than was suggested by the experiment with the tablecloth which formed our starting point. Instead of the tablecloth being rumpled over the table by an active push from the south it seems more probable that the table moved southward whereas the tablecloth was rumpled because it could not partake in the southward movement beyond a certain limit set by a resistant obstacle. The arcuate bundle of Jura folds is an exceptional phenomenon due to the remarkable coincidence of several factors. A chord like this occurred rarely — perhaps only once —in the *Symphony of the Earth*.

REFERENCES

Chapter III is adapted from three lectures. One was delivered before the Diligentia Society in The Hague, January 9th, 1948, the other two before the R. Netherlands Academy of Sciences respectively on June 26th and October 30th 1948.

The principal publications to which reference was made in the text are enumerated in the following bibliography.

ANNAHEIM, H., *Studien zur Geomorphogenese der Süd-Alpen zwischen Gotthard und Alpenrand* (Geographica Helvetiae 1, 1946, pp. 65–149).

ARBENZ, P., *Die Helvetische Region* (Guide de Geolog. de la Suisse 1934, fasc. II, pp. 96–121).

ARGAND, E., *Sur l'arc des Alpes occidentales* (Ecl. Geol. Helv. 14, 1916).

ARGAND, E., *La zone penninique* (Guide Geolog. de la Suisse, 1934, fasc. III, pp. 149–190).

AUBERT, D., *Le Jura et la tectonique d'écoulement.* Mém. Soc. Vaud. Sc. Nat. vol. 8, no. 4, 20 pp. 3 fig's, 1945.

AUBERT, D., *Le Jura,* (Geologische Rundschau, 37, 1949, pp. 2–18).

BAILEY, E. B., *Tectonic Essays mainly Alpine* (Oxford 1935).

BARBIER, R., *Les zones ultradauphinoise et subbriançonnaise entre l'Arc et l'Isère.* (Mém. Carte Géologique de France, Paris 1948).

BAUMBERGER, E., *Die Molasse des Schweizerischen Mittellandes und Jura gebietes.* (Guide Géolog. de la Suisse, fasc. I, pp. 47–75, 1934).

BEARTH, P., *Über den Zusammenhang von Mte. Rosa und Bernhard Decke.* (Eclogae Geolog. Helvetiae 43, 1934).

BEARTH, P., *Über spältalpine granitische Intrusionen in dem Monte Rosa Decke.* (Schweizer. Mineral. Petrographische Mitt. 25, Zürich 1945).

BERSIER, A., *Recherches sur la géologie et la stratigraphie du Jórat.* (Bull. d. Lab. de Géologie etc. de l'Université de Lausanne. Bull. 63, 1938).

BUXTORF, A., *Geologische Beschreibung des Weissensteintunnels und seiner Umgebung.* (Mat. Carte Géol. Suisse, n. ser., Livr. 21, 1908).

BUXTORF, A., *Prognosen und Befunde beim Hauensteinbasis-und Grenchenbergtunnel und die Bedeutung des letztern für die Geologie des Juragebirges.* (Verh. Natf. Ges. Basel 27, 1916).

BUXTORF, A., *Tafeljura-Hauensteingebiet.* (Guide Geolog. de la Suisse, Fasc. VIII, pp. 525–532, 1934).

BUXTORF, A., *Zur Altersfrage der Faltungsphasen im Kettenjura.* (Eclog. Geolog. Helvetiae, 31, p. 381, 1938).

CADISCH, J., *On some problems of Alpine Tectonics* (Experientia II, 1946).

COLLET, L. W., *The Structure of the Alps* (E. Arnold, London 1927).

DOZY, J. J., *Beitrag zur Tektonik der Bergamasker Alpen* (Leidsche Geol. Mededeelingen VII, 1935, pp. 63–84).

EARDLEY, A. J. AND WHITE, M. G., *Flysch and Molasse* (Bull. Geolog. Soc. of America 58, pp. 979–980, 1947).

ERZINGER, E., *Die Oberflächenformen der Ajoie, Berner Jura,* (Thesis, Basel 1943).

FAVRE, J. ET A. JEANNET, *Le Jura.* (Guide Geolog. de la Suisse, fasc. I, 42–56, 1934).

FRÖLICHER, H., *Geologische Beschreibung der Gegend von Escholzmatt im Entlebuch (Luzern)* (Beitr. Geol. Karte der Schweiz NF. 67 Lief. 1933).

GAGNEBIN, E., *Les idées actuelles sur la formation des Alpes.* (Actes Soc. Helvet. Sci. nat. 1942, pp. 47–53).

GLANGEAUD, L., *Gravimetrie, tectonique fine et structure profonde de la bordure externe du Jura.* (C.R. de l'Acad. des Sci. Paris 216, pp. 671–673, 1943).

GLANGEAUD, L., *Le rôle des failles dans la structure du Jura externe.* (Bull. de la Soc. d'Hist. natur. du Doubs 51, pp. 17–36, 1944).

GÜLLER, A., *Zur Geologie der südlichen Mischabel und der Monte Rosa Gruppe.* (Eclogae Geolog. Helvetiae, vol. 40, pp. 39–161, 1947).

GÜNZLER-SEIFFERT, H., *Probleme der Gebirgsbildung.* (Mitt. Naturf. Gesellsch. Bern N.F. 3, pp. 13–31, 1945).

HABICHT, K., *Geologische Untersuchungen im südlichen sanktgallisch-appenzellischen Molassegebiet.* (Beitr. Geol. Karte der Schweiz N.F. 83, pp. 1–168, 1945).

HAUG, E., *Contribution à une synthèse des Alpes Occidentales.* (Bull. Soc. Geol. de France, 4e ser., Vol. XXV, pp. 97–244, 1925).

HEIM, A., *Geologie der Schweiz.* (Tauchnitz, Leipzig Bd. I, pp. 443–704, 1919).

KNOBLAUCH, P. UND REINHARD, M. *Blatt Iorio und Erläuterungen.* (Geolog. Atlas der Schweiz 1 : 25000, Blatt 516, 1939).

KOBER, L., *Der Geologische Aufbau Osterreichs.* (Springer, Wien 1938).

KOPP, J., *Zur Tektonik der westschweizerische Molasse.* (Eclogae Geolog. Helvetae, vol. 39, no. 2, pp. 269–274, 1946).

LINIGER, H., *Geologie des Delsberger Beckens und der Umgebung von Movelier.* (Mat. Cart. Geol. Suisse, n. ser., livr. 55, 1925).

LOMBARD, A. E., *Appalachian and Alpine structures. — A comparative study.* (Bull. Americ. Assoc. of Petrol. Geolog. 32, pp. 709–744, 1948).

LUGEON, M., *Une hypothèse sur l'origine du Jura.* (Bull. d. Lab. de Géol., Min., etc. Univ. de Lausanne, Nr. 63, 1941).

MARGERIE, E. DE, *Le Jura.* (Mém. Carte Geol. de France I 1922; II 1936).

OULIANOFF, N., *Infrastructure des Alpes et tremblement de terre du 25 Janvier 1946.* (Bull. Soc. Geol. de France, 5e ser. 17, pp. 39–53, 1947).

PREISWERK, H. UND REINHARD, M., *Geologische Ubersicht über das Tessin.* (Guide Geolog. de la Suisse, fasc. III, pp. 193–205, 1934).

PHILIPP, H., *Die Stellung des Jura im alpin-saxonischen Orogen.* (Zeitschr. Deutsch. Geol. Gesellsch. 94, pp. 373–486, 1942).

SCHNEEGANS, D., *Sur l'age des failles du Jura alsacien.* (C.R. Soc. Geol. de France, pp. 24–25, 1932).

SITTER, L. U. DE, *The principle of concentric folding and the dependence of tectonical structure on original sedimentary structure.* (Proceed. Kon. Ned. Acad. v. Wetensch. Amsterdam 42, pp. 412–430, 1939).

SITTER, L. U. DE, *L'histoire géologique des Alpes Tessinoises entre Lugano et Varese.* (Leidsche Geol. Mededeelingen Dl. XI, p. 1–61, 1939).

SITTER, L. U., DE AND C. M. DE SITTER-KOOMANS, *The Geology of the Bergamasc Alps, Lombardia, Italy.* (Leidsche Geologische Mededeelingen XIV, B, 1949, pp. 1–259).

SITTER, L. U. DE, *La Géologie des Alpes Méridionales d'après les levés récentes.* (Geologie en Mijnbouw I, pp. 68-91, 1939).

SITTER, L. U. DE, *Antithesis Alps Dinarides.* (Geologie en Mijnbouw 9, p. 1–147, 1947).

SITTER, L. U., DE, AND C. M. DE SITTER-KOOMANS, *The geology of the Bergamasc Alps, Lombardia, Italy,* (Leidsche Geologische mededeelingen XIV. B. 1949, pp. 1–259.

STAUB, R., *Der Bau der Alpen.* (Beitr. z. geol. K. d. Schweiz N.F. 52, 1924).

STEINMANN, G., *über die tektonischen Beziehungen der Oberrheinischen Tiefebene zu dem nordschweizerischen Kettenjura.* (Ber. Naturf. Gesellsch. Freiburg i.Br. 6, pp. 150–159, 1892).

TERCIER, J., *Dépôts marins actuels et séries géologiques.* (Eclogae Geolog. Helvetiae, 32, pp. 47–98, 1939).

TERCIER, J., *Le problème de l'origine des Préalpes.* (Bull. Soc. Fribourgeoise des Sci. Nat., vol. 37, pp. 125–140, 1945).

UMBGROVE, J. H. F., *Origin of the Jura Mountains.* (Proc. Kon. Ned. Acad. v. Wetensch., vol. LI, Nr. 9, pp. 1049–1062, 1948).

VONDERSCHMITT, L., *Die geologischen Ergebnisse der Bohrungen von Hirtzbach bei Altkirch (Ober Elsass).* (Eclogae Geolog. Helvetiae 35, pp. 76–98, 1942).

WANNER, E., *über die Mächtigkeit der Molasseschicht.* (Vierteljahresschrift des Naturf. Gesellsch., Zürich 79, p. 244, 1934).

# Chapter IV

# DEEP FURROWS ON THE CONTINENTS
# AND IN THE DEEP-SEA

## INTRODUCTION

The subject mentioned in the title is worth being treated at greater length and in more detail than can be done to-day. As a matter of fact, the time for this communication is limited to 15 minutes which means that there is no opportunity for long meditations. For a short introduction two minutes must suffice; just enough to say that I must restrict myself seriously and that I can summarize only very briefly a few remarkable features concerning some deep furrows in the East Indies and in Burma.

What I want to make clear within the next quarter of an hour is this:

A comparison of the structural zones of the East Indies with those of Burma furnishes an important clue concerning the origin of the East Indian deep-sea troughs. For, it appears from such a comparative study that the East Indian deep-sea troughs have their counterpart in similar furrows on the Asiatic continent. These, however, were filled up entirely with sediments and were eventually raised above sea-level.

## DEEP-SEA FURROWS

Let us first look at the troughs in the western part of the East Indies. (The numerals in the following text correspond with those on Plate VI which should be consulted continually).

There are two strings of islands. Take, for example, Sumatra and the islands to the west of it. The latter continue eastward as a submarine ridge which ends to the west of Sumba island.

This non-volcanic outer arc is bounded on either side by a series of deep-sea troughs. One consists of the elongated marginal deeps. The second series of troughs extends along the internal concave side of the outer arc, that is to say between the two arcs and their submarine continuations.

Then, proceeding farther inward, follows the mountain-ridge (gean-

ticline) of western Sumatra and the southern part of Java. It is crowned
with numerous volcanoes.

Yet another series of troughs can be traced over Sumatra and Java,
along the inner side of the geanticline. These troughs were, however,
entirely filled up with sediments and subsequently raised above sea-level.
Only small areas are still below sea-level.

The outer non-volcanic row of islands is characterized by two features,
viz., (1) strong folding in the Miocene as well as during several earlier
epochs; (2) gravimetrically it is a zone of negative anomalies.

The strip of negative anomalies implies that there is a corresponding
deficiency of density in the earth's crust beneath. The only explanation

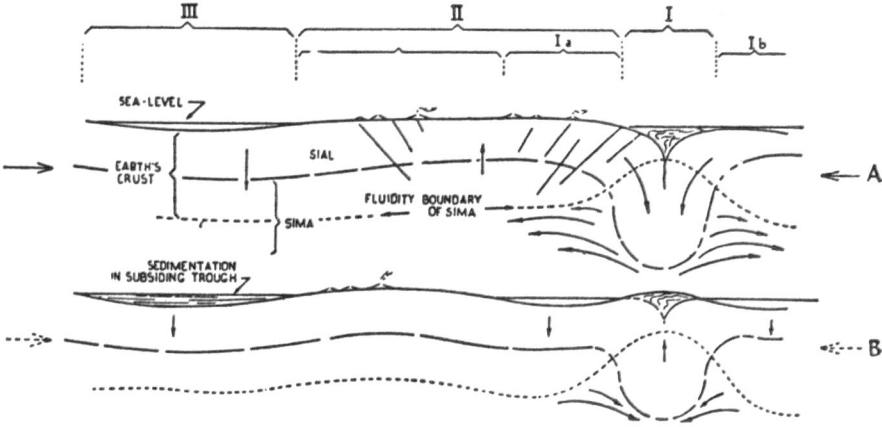

Fig. 59. Two schematic sections showing tentatively the development of structural zones in the
East Indies.

covering all the observed facts is that the upper, i.e. lighter layers of the
crust, must have been forced downward so as to form a great root of
light material. In the opinion of Vening Meinesz the root pushed and
displaced sideways the heavier material that originally built up these
deeper parts of the crust and the underlying substratum, more or less in
the way shown in fig. 59.

We may assume that after an epoch of strong compression, involving
folding of the belt and down-buckling of a sialic root, compression in the
crust decreases. This will allow the mountain-root to rise and consequently
to readjust its isostatic balance. In this process the zone of the mountain
root will probably detach itself more or less from the adjoining areas of the
earth's crust. At the same time the bordering strips must sink back.
Possibly this effect is strengthened by the inflow of subcrustal material

from the neighboring areas towards the rising zone. As a consequence of these processes a subsiding furrow will come into existence on either side of a rising mountain belt. In the East Indies these furrows (I,A and I,B) are below sea-level.

### CONTINENTAL FURROWS

As a matter of fact, the structural elements of the East Indies can be traced very clearly as far as Burma.

Evidently, the continuation of the belt of strongly negative anomalies which follows the submarine ridge south of Java and thence runs over the islands west of Sumatra appears again in Burma. Actually, Evans and Crompton found a gravity minimum coinciding with the Burmese mountain belt of the Arakan Yomas (I).

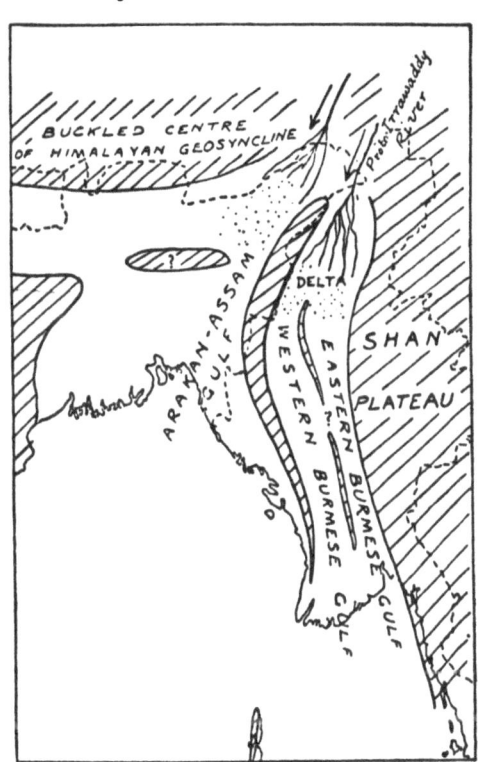

As in the East Indies, zone I is accompanied on either side by a trough. In Burma, however, the troughs are entirely filled up with sediments, ranging from Eocene to Pliocene in age. The oil-bearing strata of Assam-Arakan form the contents of a trough which can be correlated with the marginal deep-sea furrows west of Nias and the Mentawei islands and S. of the submarine ridge, south of Java (I,B).

In a similar way the Mentawei trough has its counterpart in the trough which runs east of the Arakan Yomas and west of the volcanic belt of central Burma (I, A).

Fig. 60. Geographic map of Burma in early Tertiary times. (After Dudley Stamp, from H. L. Cchibber).

It was Dudley Stamp who constructed the map reproduced in fig. 60 portraying the paleogeographic conditions of Burma during early Tertiary times. On this map the area between Arakan Yomas and Shan Plateau is subdivided into two separate sedimentation troughs by a narrow median ridge. The ridge coincides with a belt of volcanic rocks and volcanoes of Tertiary till sub-recent age. This belt (II) can be traced towards the volcanic geanticline of western Sumatra and southern Java via the volcanic islands Narcondam and Barren Island.

The Tertiary trough east of the volcanic belt of central Burma contains again a sequence of strata ranging from Eocene to Pliocene. It corresponds to similar troughs in the eastern part of Sumatra and the northern part of Java (III).

The eastern Burmese trough is bounded on the continental side by the Shan Plateau consisting of pre-Tertiary rocks which have been intensely folded (Laramide phase). We find the continuation of this Tertiary land in Malaya, Riouw Archipelago, Banka, Billiton and the western part of Borneo, including the greater part of the South China Sea and the Java Sea (zone IV).

CONCLUSION

Now let us return once more to the East Indies.

The present stage of the East Indies is shown in a tentative and schematic block-diagram (fig. 10). Filling-up of the trough behind the inner arc could occur only if the quantity of waste products from the surrounding areas equaled or surpassed the amount of room provided by subsidence of the bottom of the trough. An example is shown in the foreground of the block. The other possibility is that the rate of subsidence of the bottom was too great to be counteracted by the supply of sediments. In that case a deep-sea trough would originate and persist. This case has been drawn also in the block-diagram. Possibly the Flores deep is an example.

The different aspects of the Burmese zones, if compared to the East Indian furrows, may be explained in a similar way. The Burmese troughs received more material from the surrounding areas. Hence, they became oil-bearing areas of economic importance in strong contrast to the East Indian deep sea furrows I A and I B.

In conclusion, I must say a few words concerning the time of origin of the deep-sea troughs.

The fact that the Burmese troughs date from the Eocene does not exclude a younger age for the East Indian deep-sea furrows.

Even if we suppose that the downward movement of the deep-sea troughs dates from early Tertiary times two possible modes of origin remain open.

One possibility is that sedimentation in the East Indian sectors was too slow from the very beginning so that there always have been submarine furrows though not necessarily as deep as they are now.

The other possibility is that originally the East Indian troughs were filled up partly or wholly with sediments in much the same way as the Assam and Burmese troughs. In that case the subsiding movement of the troughs must have been accelerated in more recent times. In this process they must gradually have been transformed into to deep furrows when the subsiding movement came to surpass the rate of sedimentation.

Several data support the conclusion about the comparatively recent origin of the deep-sea areas of the East Indies as depressed continental areas. In several places the trend of the folded Miocene strata is intersected at an angle by the present coastline which at the same time is the boundary of a deep-sea trough or basin. Hence, the deep-sea areas must either have originated or they must have been rejuvenated and grown wider at least after the Miocene folding. In other places subsidence of the sea-floor of the order of some thousands of meters is also deduced from some atolls and barrier reefs. Taka Garlarang atoll, for example, rises up abruptly from a depth of over 2000 metres and is built on the side of a deeply submerged ridge running from South Celebes to the south. The steepness of the submarine slopes gives convincing evidence of their having been formed by up-growth of reef structures on a subsiding bottom. Similar conclusions are applicable to the atolls of the Tukang Besi islands, southeast Celebes. In the middle of three of the accompanying islands outcrops of Tertiary strata were found. Hence, the conclusion is again reached that the beginning of the movement dates from Pleistocene times or at the most from the uppermost Tertiary. The same conclusion can be deduced for the deep-sea areas bordering the barrier-reefs of the Togian islands, North Celebes. On the other hand, several islands in the Moluccas show signs of elevation of the order of a thousand metres and more since Upper Tertiary times. And as the subsidence of the troughs was probably related to the rising movement of the intervening strips of islands this would once more point to the comparatively recent origin or rejuvenation of the deep-sea furrows in the East Indies.

This chapter, a very short one, has been inserted in this place because it forms as it were an intermezzo between chapters III and V. It may be recalled that on our excursion across the Alps, a feature was found comparable to the situation in the East Indies and Burma viz. a mountain-belt on either side of which a subsiding trough originated when the folded chain rose up. I refer to the origin of the Lombardic and Molasse troughs after the Oligocene phase of folding of the Alps.

However, there is a striking difference between the East Indies and the Alps. The negative gravity anomalies found in the East Indies are much greater than those in the Alps. This may be due to differences in age, but the East Indian anomalies are also confined between comparatively narrower and sharper limits. In order to elucidate this question we have to proceed, in the next chapter, to some meditations on the root of the Alps.

## REFERENCES

The contents of this chapter correspond largely to a communication read by the present author before the International Geological Congress in London, August 1948. Apart from a more detailed publication, which will appear in the Proceedings of the Congress under the title "Origin of the deep-sea troughs of the East Indies", the following publications may be cited for fuller reference.

CCHIBBER, H. L., *The Geology of Burma*, (1934).

EVANS, P. AND CROMPTON, W., *Geological factors in gravity interpretation illustrated by evidence from India and Burma*. (Quarterly Journ. Geolog. Soc. London CII, 1946, pp. 211–247).

SALE, H. M. AND EVANS, P., *The geology of the Assam-Arakan Oil Region (India and Burma)*. (Geolog. Magaz. LXXVI, 1940, pp. 337–363).

STAMP, L. D., *An outline of the Tertiary Geology of Burma*. (Geolog. Magaz. LIX, 1922, pp. 481–501).

TIPPER, G. H., *The Geology of the Andaman Islands with reference to the Nicobares*. (Mem. Geol. Surv. India 13, 1911, pt. 4, p 1).

UMBGROVE, J. H. F., *Structural history of the East Indies*. (Cambridge University Press, 1949).

UMBGROVE, J. H. F., *Coral reefs of the East Indies*. (Bull. Geolog. Soc. of America 58, 1947, pp. 729–778).

VENING MEINESZ, F. A., KUENEN, PH. H., AND UMBGROVE, J. H. F., *Gravity Expeditions at Sea* (II, 1934).

Plate VI

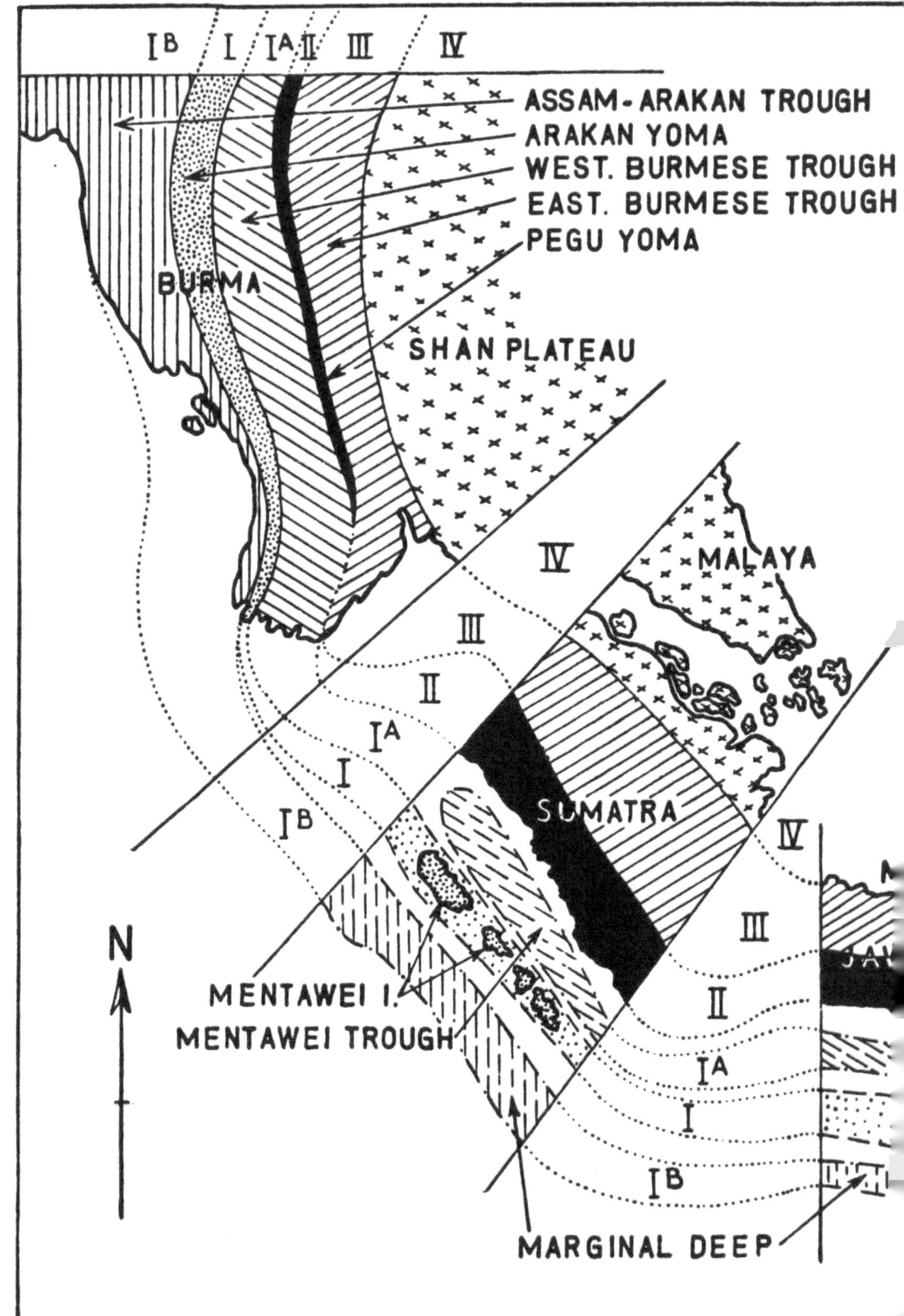

ASSAM-ARAKAN TROUGH
ARAKAN YOMA
WEST. BURMESE TROUGH
EAST. BURMESE TROUGH
PEGU YOMA

Structural zones in different s

I {
Zone of strong Laramide folding. Miocene epochs of compression, weak in Burma, strong in other sectors. Zone of strong negative anomalies of gravity.

Submarine continuation of the same zone.

A {
Inner trough, accompanying zone I, filled up with Tertiary sediments in Burma sector.

Submarine inner trough.

B {
Outer trough, accompanying zone I, filled up with Tertiary sediments in Burma sector.

Submarine outer trough, marginal deep.

II
Geanticlines, Tertiary up to recent volcanism. Miocene diastrophism moderate.

Sumba Isl. Upper Miocene folding. Tertiary volcanism. No recent volcanoes. Interruption of zones $I, I^A, I^B$ and of zone of strong negative anomalies of gravity.

I {
Tertiary sedimentation troughs; moderate folding towards the end of the Pliocene.

Submarine continuation of the same troughs.

I
Regions above sea-level in Tertiary and Pleistocene times.

SURA   MADURA STRAITS   IV

FLORES DEEP

BALI   III

II

SUMBAWA   FLORES

I^A

SUMBA   TIMOR

I

I^B

SUBMARINE RIDGE

III

II

I^A

I

I^B

CHAPTER V

# THE ROOT OF A MOUNTAIN-CHAIN

## INTRODUCTION

Probably you will feel it as a queer situation that a man living in a flat country, a few metres below the level of the sea, makes so bold as to lecture in Geneva on the deeper structure of a lofty mountain-chain.

The first excuse for doing so is that the Chairman of the Geological Section kindly asked me to deliver this address and that I thankfully accepted his much appreciated invitation. For, on several occasions when travelling in this beautiful and attractive country, I not only admired the grandeur of the magnificent mountain scenery but I also became absorbed in some of its fascinating problems. The opportunity to unfold a theory on the root of the Alps before an audience of Alpine experts formed a thrilling prospect. To be sure, at this very moment it gives me more or less the feeling one must experience when entering a lion's den.

Perhaps I may put forward a second excuse by mentioning that, for several years, I worked on problems related to the deeper structure of the East Indies. Now, the East-Indian island-arcs form, so to speak, the eastern continuation of the Alps. Hence, I had some reason for plumbing also this part of the world. By way of introduction, let me try to elucidate this point in a few words.

You probably know that Vening Meinesz found a zone of strongly negative gravity anomalies in the East Indies. To explain these anomalies he assumed a thickening of the crust by a downward root which takes the place of the denser material in deeper realms and in this way causes the lack of gravity attraction. In other words, he suggested that under the action of great horizontal stresses the crust buckled downward along the axis of the belt so as to form a great downward protuberance. The folding and overthrusting of the surface strata, as found by the geologist, are considered as an accompanying feature of this great phenomenon. Geologically speaking this means that a process of underthrusting towards the central zone was of major importance. All geological data known from the outer arc fit Vening Meinesz'theory remarkably well and, according to my opinion, furnish a splendid confirmation of it. Plate VI and the block-diagram of fig. 10 are schematic representations of the East Indian island arcs. The strip of negative anomalies follows the outer arc. The latter indeed shows the features of strong compression and overthrusting which we might expect to find in such a zone. The question now arises whether similar phenomena are known to the west. And so we arrive at the structural problems of the Swiss Alps. Fifteen years ago Vening Meinesz suggested the structure of the Alps might be explained by a similar process of down-buckling of the crust.

At that time, however, the prevailing theory among Swiss geologists was very different. Probably even to-day most of you adhere to a theory which is different in its fundamental aspects. I mean the well-known synthesis of Argand and its elaboration by Staub.

A prominent feature in Argand's tectonic synthesis of the Alps is the suggested northward overthrusting of the southern "hinterland" of the old Tethys geosyncline. In its most extreme expression this theory was formulated as overthrusting of Africa over Europe, the crystalline cores of the Austro-Alpine nappes being regarded as frontal overthrust parts of the hinterland, and the Pre-Alps as a small part of Africa resting on Europe (fig. 53, p. 78).

I venture to suggest, however, that progressive underthrusting from both the foreland and "hinterland" towards the centre of the belt was the main process in the structural history of the Alps.

Geological as well as geophysical evidence support this point of view. The geological motives were amply discussed in chapter III. Hence this chapter will be restricted to the geophysical side of the problem.

### GEOPHYSICAL EVIDENCE

Suppose progressive underthrusting towards the mountain-chain from both the northern and southern foreland has taken place. Then, evidently, the two girdles of northern and southern massifs (see Pl. VII) were zones of very high friction and stress, especially their inner sides where they are bounded by the huge Pennine nappes and their roots. Probably the same strips were predestined to become major zones of movement during the subsequent process of restoration of isostatic equilibrium of the Alps. This suggestion is supported by the distribution of seismic belts. In the western Alps Rothé found two belts of seismic activity (fig. 61). The northern belt corresponds to the inner side of the zone of "Central

Massifs". The massifs themselves are practically aseismic. The conspicuous accumulation of epicentres between the massifs apparently means that similar tectonic elements are present at a lower level where they are buried in the intervening area of axial depression. The southern seismic belt corresponds to the boundary between the Pennine roots and the Ivrea zone which is characterized geologically as a steeply dipping zone of mylonitization and great faults (Ivrea, Tonale, Orobic and Insubric belts).

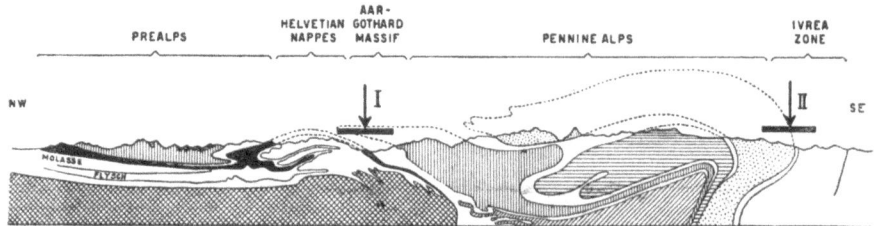

Fig. 61. Position of seismic belts (I and II) in the Alps. (After Rothé).

Another consequence of the buckling theory leads to still deeper problems in a litteral sense. For, if underthrusting of the foreland towards the Alps took place from both the northern and the southern sides a sialic root of large dimensions must have been forced downward under the present mountain-chain.

Indeed, gravity anomalies found in the Alps clearly demonstrate the existence of a root of comparatively light material below the mountain-chain.

Niethammer's map of Bouguer anomalies shows isanomaly curves roughly parallel to the general trend of the mountain-chain. The anomalies gradually increase from zero along the northern margin of the Jura Mountains up to about -150 in the Pennine Alps, whence they decrease again to zero when proceeding towards the southern margin of the Alps.

Salonen's sections of Bouguer anomalies constructed at right angles to the trend of the Swiss mountains reveal an additional steepening below the Pennine Alps apart from the general increase of the anomalies towards the centre.

The same feature is especially clear in profiles of the Eastern Alps constructed by Holopainen with the aid of a modified Bouguer reduction. The same phenomenon appears in all his curves of isostatic or Airy anomalies based on various assumptions of the thickness of the crust (T) and the degree of regional compensation (R). One of Holopainen's profiles is reproduced in fig. 62.

The general conclusion deduced from studying these anomaly curves is: (1) the presence of a broad root of light material below the Alps and the adjacent regions gradually increasing from the northern and southern boundaries towards the central belt of the Alps, (2) an additional root of light material below the central belt.

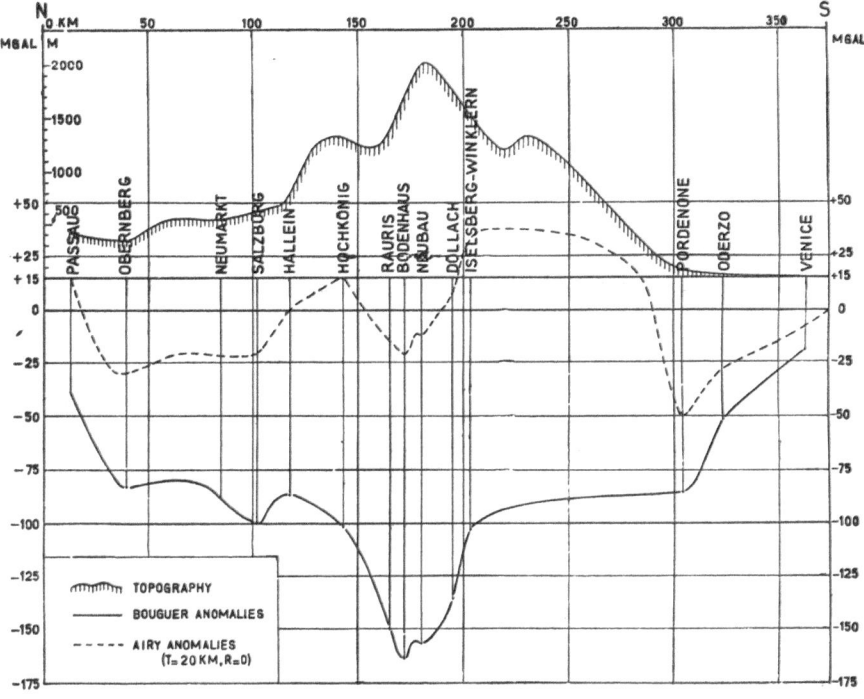

Fig. 62. Gravity profile across the Eastern Alps. (After Holopainen).

From a purely physical point of view numerous, not to say an infinite number of suppositions about the position of the light masses might be put forward.

The question now arises what the most probable distribution is. The relatively sharp peaks of the curves furnish a means of limiting the number of suppositions considerably. For, they make clear that most probably the disturbing masses occur at comparatively shallow levels. A comparison with the geological features at the surface furnishes a further means of arriving at an explanation, which in some cases seems highly probable. Thus, for example, one sharp down-bending of the isostatic anomaly curves (fig. 62) corresponds exactly to the site of the Molasse trough and a similar down-bending corresponds to the Lombardic trough. Evidently, the northern and southern strips of negative anomalies are due to the presence of the Molasse and Lombardic troughs. Holopainen thinks the negative anomaly is due either to the prism of young and light sediments with density of about 2.37 or to the non-equilibrium of the area or to a combination of both factors.

It is difficult to locate the disturbing mass which causes the negative bulge below the central belt. Tracing the structural history of the Alps (Chapter III) we found that underthrusting towards the central belt probably was a dominating process. Hence, the theory of a root of light material was postulated as a logical and necessary consequence. However, geological data cannot possibly furnish any evidence about the shape and dimensions of the root.

Holopainen's attempt at interpretation of the gravity data is based on several uncertain premisses.

His theory starts from two fundamental assumptions. One is a "normal anomaly" of + 15 milligal for the whole area of western Europe, as based on a formula accepted by Heiskanen in 1938. The meaning of this positive anomaly is a mystery. Moreover, a different value may result if future investigations enable us to compute the "normal anomaly" on more numerous and more accurate data. The weight of this uncertainty will be clear if we realize that a second assumption is intimately connected with it, viz. the supposed thickness of 20 km of the earth's crust for zero elevation ($T = 20$). For the sake of simplicity Holopainen accepts the following model of the earth's crust. A crust 20 kilometres thick and with a density of 2.67 floating on a substratum with density 3.27. In his opinion these figures represent the most probable assumption (fig. 63). Of course, the negative values partly depend on the accepted "normal anomaly". If the normal anomaly were to prove less than + 15 m.gal the result would be a corresponding increase of the thickness of the crust in Holopainen's model.

Moreover, it seems more probable that the crust consists of several layers of varying density. For evident reasons the simplification introduced by Holopainen in his one layer model is another factor that affects the conclusions based on it. Fig. 63 clearly shows what his conclusions are. The lower surface of the crust is supposed to occur at a depth of 20 kilometres for zero elevation. An additional root of about 7,5 to 10 kilometres is drawn below the central Alps corresponding to the mean elevation of about 2 kilometres. Hence, a value of about 30 km would be probable for the thickness of the earth's crust under the mountain range. An additional root, called downward bulge, increases the thickness under the central belt by an amount of about 10 km over a north to south distance of about 20 km. According to Holopainen the downward bulge originated by a process of down-buckling as suggested by Vening Meinesz for the belt of strongly negative anomalies in the East Indies. It would be premature, however, to consider this model as a picture of the real situation.

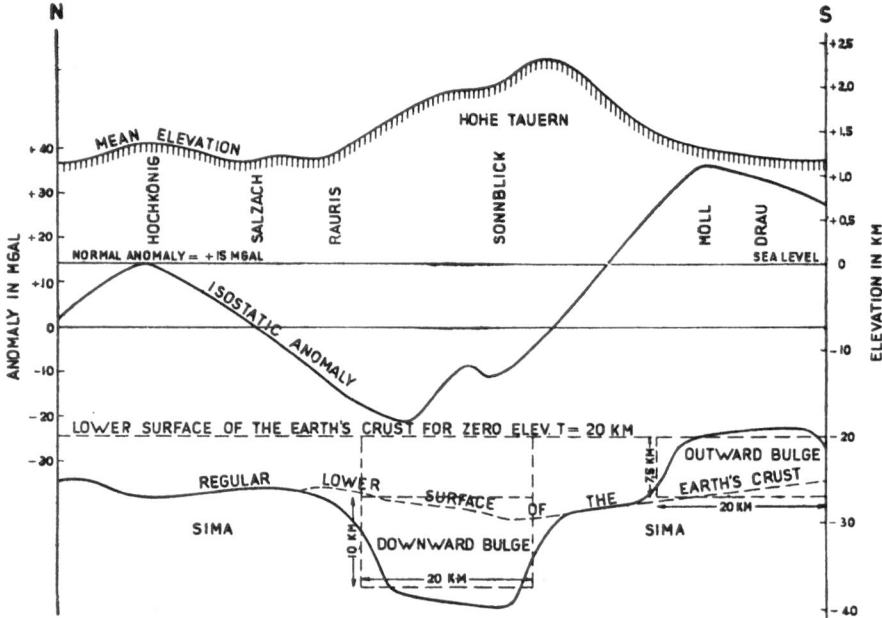

Fig. 63. Interpretation of isostatic anomalies in the Eastern Alps. (After Holopainen).

Fig. 64 represents a crustal model built up of a granitic layer and intermediate layers above the substratum. The density of the intermediate layers is supposed to be greater than of the granitic layer and to be surpassed by the density of the substratum, though no special values will be introduced. The geological profile of fig. 64 is based on Staub's profile no. 4, which coincides with Holopainen's profiles II and IIa (our fig. 62 and 63). Sedimentaries are marked by dots, crystalline cores of nappes are left white. Evidently, the negative anomalies observed at the surface result from the combined influences of the sediments indicated by s and the roots $r_1$ and $r_2$. Probably the highly elevated pile of comparatively light sediments indicated by s in the geological section has a marked influence on the gravity curve. It is suggested that after its substraction from the total effect the resulting curve would coincide approximately with the dot-dash line. Hence, the remaining negative must be due to the roots r, $r_1$ and $r_2$.

The downward bulge r is narrower than $r_1$ which in turn is narrower than $r_2$, but it is impossible to decide on the real magnitudes of the

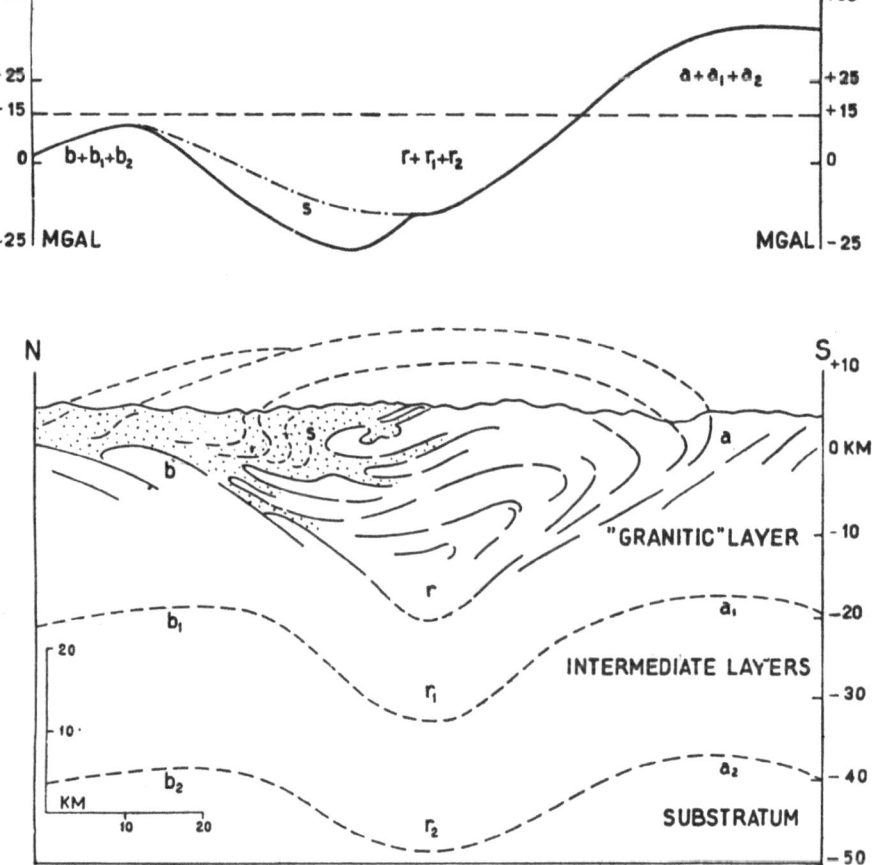

Fig. 64. Attempt at interpretation of isostatic anomalies in the Eastern Alps.

respective downward bulges, because data on the thicknesses of the crustal layers and their specific densities are still too few and of a too uncertain character.

*Mutatis mutandis* the same holds good regarding the outward bulges a, $a_1$, $a_2$ and b, $b_1$, $b_2$ which correspond at the surface respectively to the zones of central massifs and southern basement rocks.

Possibly the positive belts are due to the higher level to which deeper and denser rocks rose automatically when they had to follow the upward movement of the northern and southern belts of basement rocks to their present high situation. Of course, other factors may have been of additional importance during the process of restoration of isostatic equilibrium in the Pleistocene. Instrusion of basic magma may be responsible for the relatively high positive anomalies in some areas, as suggested by E. Niggli for the region near Lake Maggiore.

A geological conclusion which seems to be warranted and to be sustained by gravimetric and seismic data is that during the long history of the Alps the crust became thickened due to progressive underthrusting towards the central belt, and that part of the thickened crust formed a central downward bulge, though it remains uncertain which crustal layer formed the main part of the root. Moreover, it seems highly probable in such a process that the phenomenon of underthrusting gradually migrated in a direction from the centre towards the "foreland" on either side of the chain.

Considering the structural history of the Alps one notices an outward progression of tectonic action resulting in the addition of more and more structural elements. The complicated system of Mesozoic troughs and intervening ridges in the Pennine region was bounded on one side by the deeply subsiding Helvetian trough on the other side by the sedimentation troughs of the Southern Alps.

After the Oligocene paroxysm two new troughs came into existence, one further to the north (the Molasse trough) one to the south (the Lombardic trough). Probably the subsidence of these troughs kept pace with the rising movement of the folded belt in between them, due to a cause to effect relation. In the meantime denudation products from the Alps filled up the subsiding troughs on either side of the rising mountain-chain.

Due to a renewed compression towards the end of the Miocene Tortonian strata of the Lombardic trough were overrun by Triassic rocks of the Bergamasc Alps, and the Helvetian nappes came to rest on Molasse deposits. Moreover, differential movements along planes separating wedge-shaped parts of the basement caused the southern boundary zone of the Molasse trough, including the frontal part of the Helvetian nappes, to become tilted and adjusted into their present position.

Usually the crystalline wedges of basement rocks have been regarded as upthrust or even squeezed out masses due to pressure from the Pennine nappes. Though it is not denied that such a process may have played a part of some importance it should be granted that underthrusting towards the Alps might also be the main factor. This holds even for intricate situations such as those represented in the Windgälle (fig. 2) and Jungfrau (Pl. IV). Moreover, the elevated position of the basement rocks is due to the same factor that was responsible for the great altitude of the Pennine nappes, viz. a subsequent isostatic rise of the mountain-chain.

During the diastrophic phase of the Alps which ended in the Oligocene the region of the present Jura Mountains was affected by differential movements of basement blocks. The movements caused a roughly NE–SW pattern of faults and short anticlines as well as a number of small faults and intervening grabens in a longitudinal direction. However, the dominating pattern of longitudinal Jura folds originated with the Upper Miocene phase. Apparently, renewed Alpward underthrusting of the basement was an important factor in the formation of these folds, as was explained at greater length in chapter III.

Possibly several problematic questions will be solved if more seismic data become available. What is known from seismic evidence about the thicknesses of the crustal layers in Europe was summarized by Gutenberg in 1943. Though these data are too few for furnishing a final solution of several problems the seismological evidence is important in more than one respect. In the first place Gutenberg's conclusion is in agreement with the geological and gravimetric evidence. "For Europe", he writes, "the maximum depth of the Mohorovicić layer is undoubtedly under the region of the Alps which means that the Alps have a root".

After this conclusion Gutenberg continues: "This root is mainly a result of a greater thickness of the uppermost (granitic) layer".

The total thickness of the upper layer under the central belt of the Alps consists of two parts, viz. (1) the crystalline cores of nappes and (2) a downward bulge. On the other hand the thickness of the intermediate layers under the Alps (20–25 km) would appear to be of the same order as the thickness of the intermediate layers under the foreland. Still, it is self-evident that during the process of down-buckling the intermediate layers must have formed an additional downward bulge similar to that of the upper layer. Therefore, it appears logical to conclude that the root of the intermediate layers — at any rate the bulk of it — has disappeared by melting and spreading in the substratum.

## THE AMOUNT OF CRUSTAL SHORTENING

The same conclusion is reached from a consideration of the shortening as revealed by the Alpine folds.

Unrolling of the Alpine nappes demonstrates a shortening of the whole chain by a considerable amount. According to Cadisch an original width of 630 km has been reduced to the present width of 150 km in post-Carboniferous times. Other estimates made by various authors vary between 200 km and more than 1000 km shortening. Sonder, however, who published a critical study on this subject, agrees that 200 km is the minimum estimate allowable. Evidently, the crystalline basement must have suffered the same shortening as the superstructure, because autochthonous sediments are directly connected with the crystalline basement in the north (Aar massif) as well as in the south (Lugano-Lombardic region). The total mass of basement rocks incorporated in massifs and crystalline cores of nappes is far from sufficient to explain a crustal shortening of 200 kilometres even if the original crust was comparatively thin. Hence, the basement must have slid downward under the Alps so as to form a sialic mountain-root.

It is worth-while to deduce the amount of basement rocks in the Alps before and after the crustal shortening of the Alpine cycle from a cross-section of 150 km of the present mountain-chain. Let us assume a crustal shortening of 200 km and a crustal thickness of 30 km at the beginning of the cycle in Triassic times.

From these minimum estimates it follows that the original profile ought to show a total of crustal material in the order of $30 \times 350$ km$^2$ = 10500 km$^2$. During the process of crustal shortening part of the crystalline basement became incorporated in massifs and the crystalline cores of nappes. In the profile represented by fig. 64 this amounts to 1500 km$^2$ above the zero line. This is a maximum estimate and it includes the part that disappeared by erosion. Allowing for a thickening of the crust under the whole area of the Alps in the order of 10 km below the zero line the total amount of crustal material now present, with the exception of the downward bulge, would amount to $1500 + (40 \times 150) = 7500$ km$^2$. Therefore, a downward bulge with a profile in the order of 10500–7500 = 3000 km$^2$ must have formed during the several epochs of compression of· the Alpine cycle. The figures chosen are unfavourable for finding a large downward bulge. Even if we would allow for still more unfavourable assumptions it seems inevitable to conclude that during the Alpine cycle crustal material formed a downward bulge which for the greater part has spread in the substratum. The only means of escaping this conclusion would be to start with a much thinner pre-Alpine crust and to allow for a much greater thickening of the crust under the whole area of the Alps. For the time being these seem very improbable assumptions. As a consequence of the process of crustal shortening Holmes is inclined to conclude that the continents must have grown progressively thicker and covered an ever smaller area in the course of geological history. As suggested by Holmes this would involve a progressive increase of the rate of denudation and geosynclinal sedimentation as well as a progressive speeding up of orogenic processes. The theory of a remarkable acceleration of these phenomena, representing a genuine departure from the theory of uniformitarianism has been advocated by several authors. It is in good agreement with Holmes' geological time curve based on the most probable ages of radioactive minerals and the maximum thicknesses of the geological systems. However, the effect of the acceleration must not be overrated. For as far as can be ascertained the major cycles of diastrophism do not display a marked speeding up of their rhythm. As a matter of fact, minor cycles plotted on the time scale show an increasing frequency but probably this phenomenon is due to the fact that unravelling the earth's structural history becomes ever more difficult the farther we try to penetrate into the past! Moreover, one should not forget that after a phase of diastrophism, when the mountain belt regains isostatic equilibrium, part of the detritus is transported to the deep-sea and is forever lost from the continents. This amount should be taken into account when estimating the progressive thickening of the continents.

One of the most baffling problems of earth science is to find the motor of the deep-seated processes which cause a rhythmic shortening of the continental sectors of the earth's crust. A vast increase of geological and geophysical data is needed before these problems can be attacked without entering into the realm of mere speculation.

## COMPARISON WITH OTHER MOUNTAIN-ROOTS

In a far distant future one may expect the upper roots (r and $r_1$) to disappear due to the combined effect of continued denudation at the surface and isostatic rise of the root.

In this connection it is interesting to compare the seismic evidence found in the Sierra Nevada. Gutenberg said: "All results available indicate that the root of the Sierra Nevada is due rather to an increase in the thickness of the deeper intermediate layers than in the thickness of the uppermost (granitic) layer". However, the Sierra is much older than the Alps.

Fig. 65 shows the granitic layer without an appreciable root as contrasted to the large and broad root of the intermediate layers. The absence

of a "granitic" root is what one might expect in a structure of great age,
where there has been ample time for an original root to disappear by the
combined effect of isostatic rise and denudation. The presence of a root
of the intermediate layers is an unexpected feature if it is true that it has
already disappeared in the Alps.

If more seismic data confirm the marked difference between the Alps
and the Sierra Nevada the cause of the different features must be sought

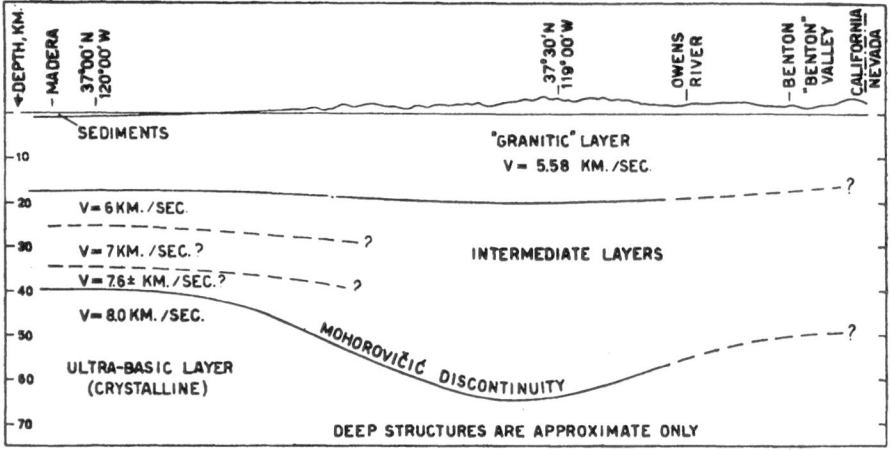

Fig. 65. Crustal layers in the Sierra Nevada. (After Gutenberg).

in fundamental differences in the structural history of these mountain-
chains.

For the time being, however, we must wait for more reliable data.
As to the Alps Gutenberg wrote: "More and better data on the velocities
in the intermediate layers are necessary to find out how far an increase in
the thickness of the intermediate layers contributes to the result".

Moreover, one may expect the boundaries between the crustal layers
to be very irregular, especially under a region like the Alps. Therefore,
many more data are needed before one can construct a satisfactory picture
of the thickness and distribution of crustal layers in the Alps.

Apparently, the processes involved in the formation of the root of the
Alps were similar to those in the East Indies in so far as a central root
originated and was perhaps rejuvenated eventually. But they were different inasmuch as the Alpine root grew ever broader during subsequent
phases of diastrophism.

Perhaps this also explains why the negative anomalies of the central
belt of the Alps are much smaller than those of the negative belt of the
East Indies.

SUMMARY

A tentative interpretation, taking into account the available geological and geophysical data is given in the tectonogram, Plate VII.

The left-hand part of the tectonogram shows the southern part of the Rhine graben (R) and Black Forest massif (BF). South of them the Jura Mountains are represented schematically. The continuation of the section through the Molasse trough (Mo) passes approximately along the famous "Axenstrasse". This part of the tectonogram, showing the structure of the so-called Helvetian Alps and remnants of the "Klippe" nappe (M) is largely based on a well-known block-diagram by Arbenz (fig. 2). The northern and southern massifs (A and Iv) as well as the Pennine nappes (I–VI, Ad, T, and Su) are drawn in a very schematic manner, as is the whole block-diagram. For the sake of clearness no "schistes lustrés" have been drawn between the Pennine nappes on the main block in the centre. They are shown, however, in the foreground block which is largely based on Argand and Bearth. The profile on the right-hand part of the tectonogram passing through the Orobic zone (Or) and the Lombardic Alps (L), is based on sections published by Dozy and de Sitter. South of them follows the Lombardic trough (Lo).

The Jura folds show a crustal shortening in the order of 10 kilometres, i.e. about 25 percent. However, it follows from the above considerations that this does not imply a proportional amount of shortening for the whole Alps during the Upper Miocene phase of compression. For, probably the greater part of the compression was taken up by the outer zones of the Alps and their foreland.

The question how the assymmetrical structure of the Alps originated above a symmetrical root was ably discussed by Bucher. According to his opinion the asymmetry results from the arcuate shape of the geosyncline. Folds and thrust-planes will show overthrusting mainly towards the convex side of the arc because movements meet less resistance towards the convex side than towards the concave side. Asymmetry is the rule, even in the rectilinear part of a geosyncline.

The situation of the young plutonic bodies along the Insubric zone or in its northern vicinity as well as their absence along the boundary of the belt of northern or so-called central massifs may also be due to the assymmetry. For, the southern zone of basement rocks is much steeper than the central massifs and so it facilitated the ascent of plutonic processes from the mountain root.

## REFERENCES

Chapter V is adapted from an address to the geological section of the *Société pour l'avancement des Sciences* at Geneva, July 1948. In the same year a communication on the same subject was given at the R. Netherlands Academy of Sciences at Amsterdam, while a publication appeared in the Proceedings of the Academy for the greater part similar to chapter V. The writer feels sincerely indebted to Professor Vening Meinesz for his elucidating and stimulating discussion of the gravity anomalies of the Alps.

Apart from the publications listed below, the bibliography at the end of chapter III should also be consulted.

BUCHER, W. H., *The Deformation of the Earth's crust.* (Princeton Univ. Press 1933).

CADISCH, J., *On some problems of Alpine Tectonics* (Experientia II, 1946).

GASSMANN, F. UND PROSEN, D., *Zur Interpretation des Schwere-defizites in den Schweizer Alpen.* (Eclog. Geolog. Helvetiae 41, 1948, pp. 135–140).

GUTENBERG, B., *Seismological evidence for roots of mountains.* (Bull. Geol. Soc. America 54, 1943, pp. 473–498).

HOLMES, A., *The construction of a geological time-scale.* (Transact. Geol. Soc. Glascow 21, part 1, 1947, pp. 117–152).

HOLOPAINEN, P. E., *On the Gravity Field and the isostatic structure of the Earth's crust in the East Alps.* (Ann. Acad. Sci. Fennicae, Serie A, III, Helsinki 1947).

HORTON, C. W., *Gravity anomalies due to extensive sedimentary beds.* (Bull. Geol. Soc. America 55, 1944, pp. 1217–1228).

JEFFREYS, H., *The Earth.* (Cambridge Univ. Press, Sec. Editt. 1929, pp. 295–296).

LEHNER, M., *Beitrage zur Untersuchung des isostatischen Kompensation des Schweizerischen Gebirgsmassen.* (Verh. Naturf. Gesellschaft, Basel, Bd. XLI, 1930).

NIETHAMMER, TH., *Die Schwerebestimmungen der Schweiz. Geol. Kommission.* (Verh. Schweiz. Naturforsch. Gesellsch. 1921).

NIGGLI, E., *Über den Zusammenhang zwischen den positiven Schwereanomalien am Südfuss der Westalpen und der Gesteinszone von Ivrea.* (Eclog. Geol. Helv. 39, 1946, pp. 211–228).

REICH, H., SCHULZE, G. A., und FÖRTSCH, D., *Das geophysikalische Ergebnis des Sprengung von Haslach im südlichen Schwarzwald.* (Geolog. Rundschau 36, 1948, pp. 86–96).

ROTHÉ, J. P., *La seismicité des Alpes occidentales.* (Bull. Soc. Géol. de France, 5e ser. II, 1942, pp. 295–320).

SALONEN, E., *Über die Erdkrustendecke und die isostatischen Kompensation in den Schweizer Alpen* (Annal. Acad. Sci. Fennice, Serie A, vol. 37, no. 3, Helsinki 1932).

SONDER, R. A., *Über das Ausmass des Alpinen Krustenzusammenschubs.* (Eclog. Geolog. Helvetiae 33, pp. 353–362, 1940).

STAUB, R., *Der Bau der Alpen.* (Beitr. Geolog. Karte der Schweiz, N.F. 52 Lief. 1924).

UMBGROVE, J. H. F., *The root of the Alps.* (Proceed. Kon. Nederl. Acad. v. Wetenschappen, LI, 1948, pp. 761–775).

VENING MEINESZ, F. A., UMBGROVE, J. H. F., and KUENEN, PH. H., Gravity Expeditions at Sea (II, 1934, pp. 129–131).

WANNER, E., *Die Erdbebenherde in der Umgebung von Zürich.* (Eclog. Geologicae Helvetiae 38, Nr. 1, 1945).

WANNER, E., *Über den Tiefgang der Alpenfaltung.* (Eclog. Geolog. Helvetiae 41, 1948, pp. 125–135).

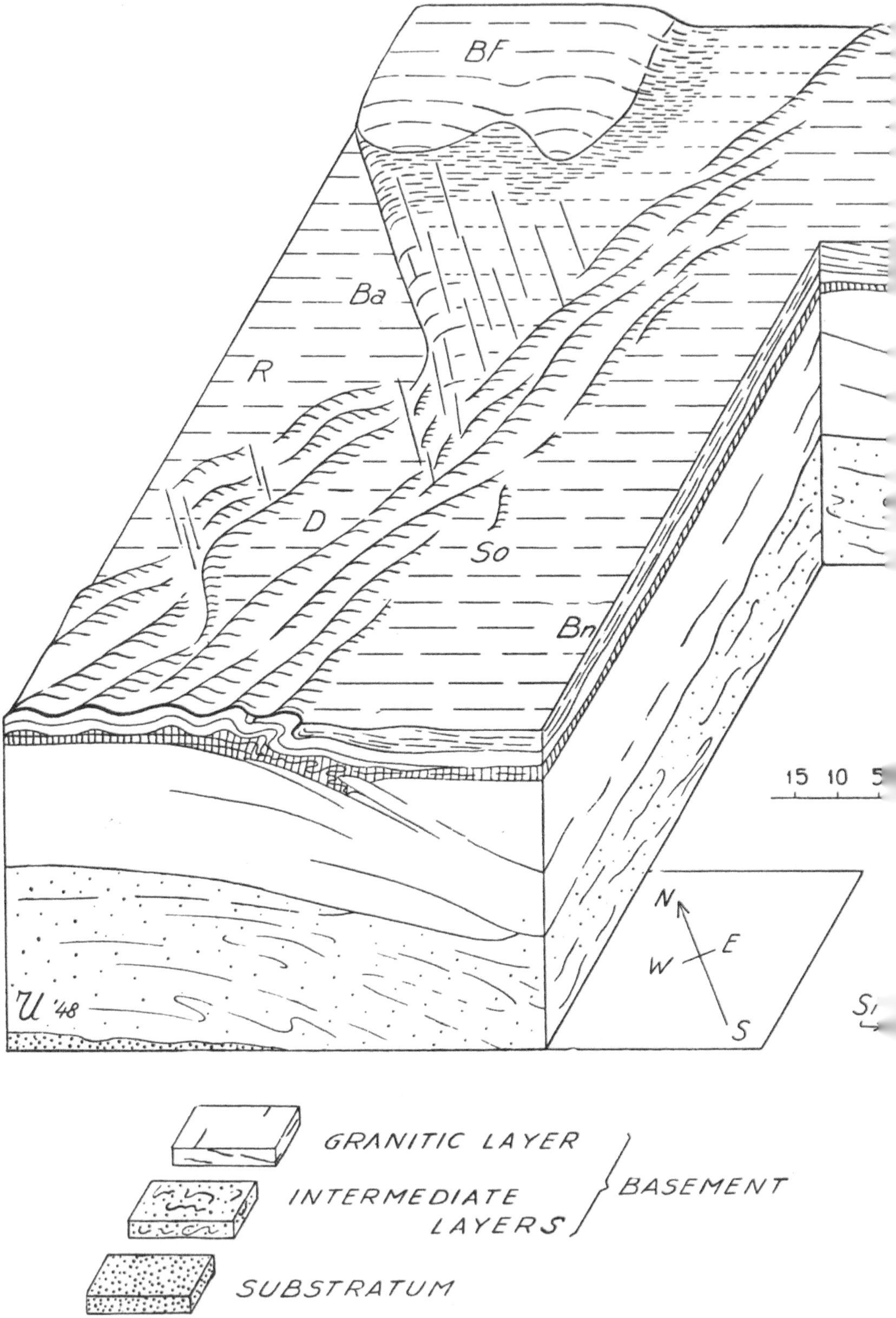

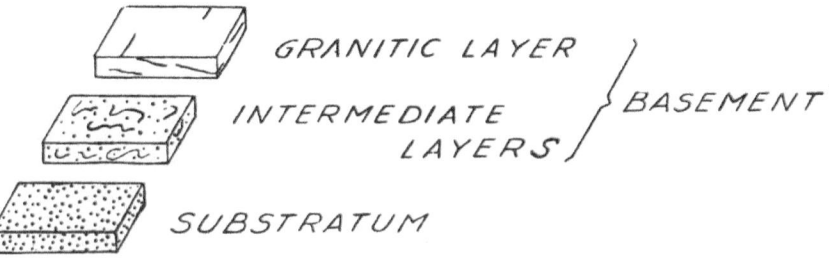

BF

Ba

R

D

So

Bn

15 10 5

N
E
W
S

𝒰 '48

S.

GRANITIC LAYER ⎫
⎬ BASEMENT
INTERMEDIATE ⎭
LAYERS

SUBSTRATUM

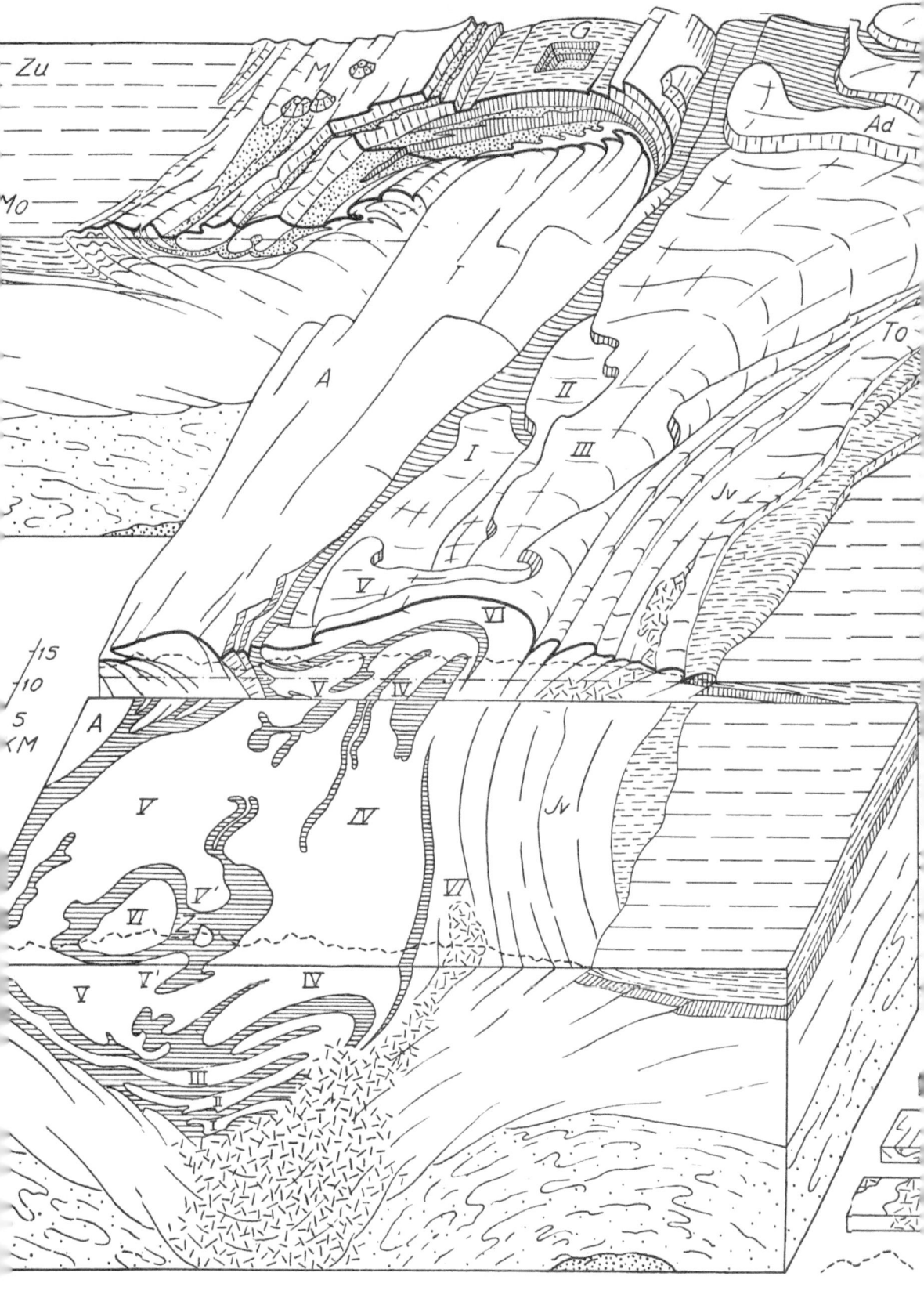

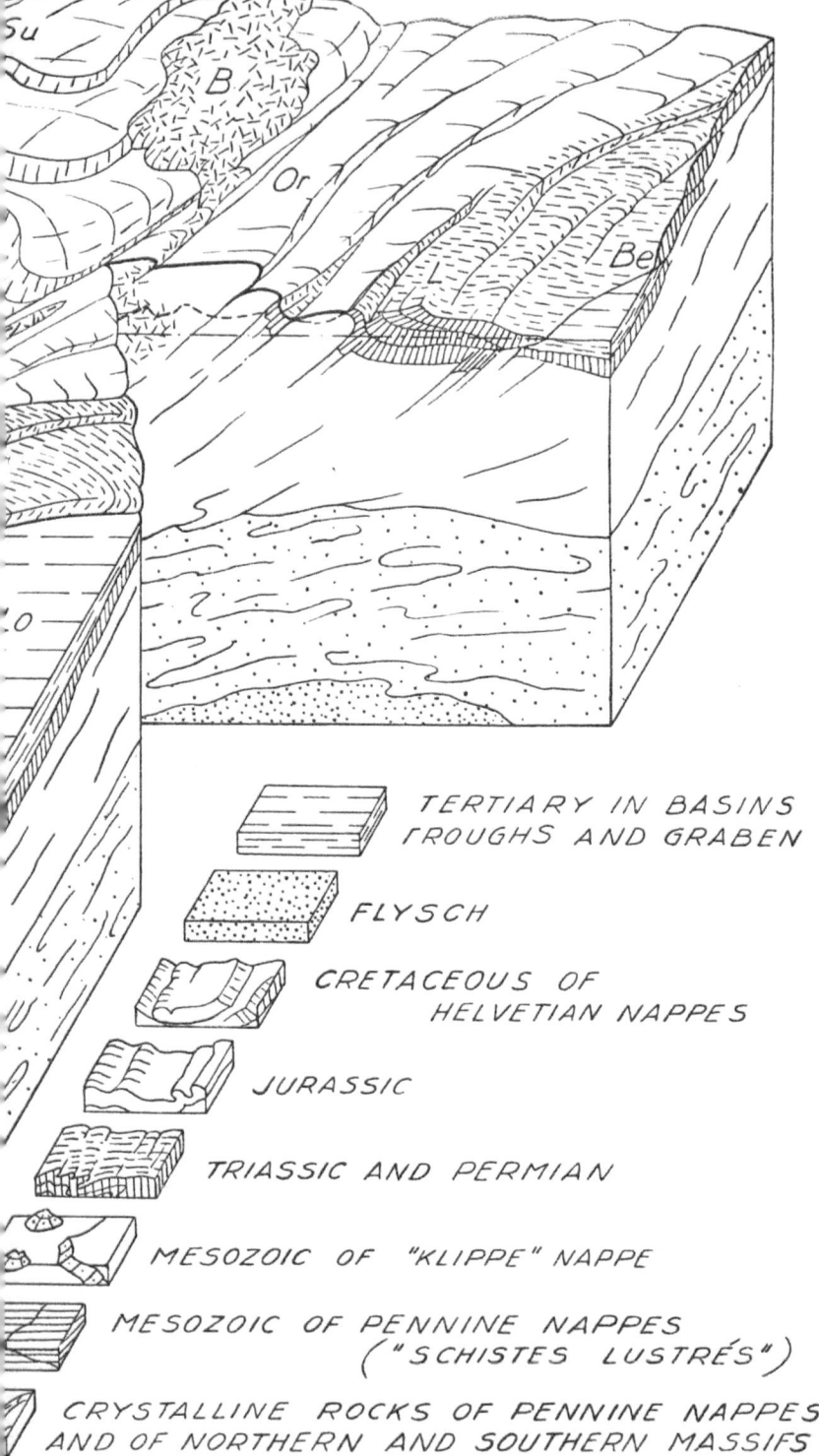

Su

B

Or

L    Be

o

TERTIARY IN BASINS
TROUGHS AND GRABEN

FLYSCH

CRETACEOUS OF
    HELVETIAN NAPPES

JURASSIC

TRIASSIC AND PERMIAN

MESOZOIC OF "KLIPPE" NAPPE

MESOZOIC OF PENNINE NAPPES
        ("SCHISTES LUSTRÉS")

CRYSTALLINE ROCKS OF PENNINE NAPPES
AND OF NORTHERN AND SOUTHERN MASSIFS

TERTIARY GRANITES, TONALITES, ETC.

POGRAPHIC PROFILE

Plate VII

## TECTONOGRAM OF PART OF THE SWISS ALPS.

A, Aar-Gothard massif; Ad, Adulla nappe; B, Bergell granite massif; Ba, Basle; Be, Bergamo; Bf, Black Forest massif; Bn, Berne; D, Delémont basin; G, tectonic window of Glarus; M, Mythen and Rotenfluh "klippe"; Mo, Molasse trough; Or, Orobic (Insubric) zone; R, Rhine graben; So, Solothurn; Si, Sion; Su, Suretta nappe; T, Tambo nappe; To, Tonale zone; Z, Zermatt; Zu, Zürich.

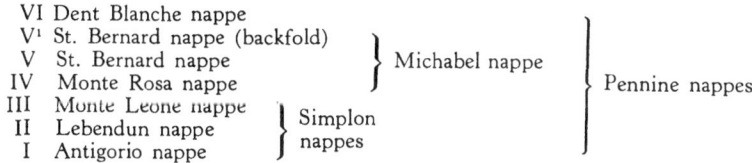

VI Dent Blanche nappe
V¹ St. Bernard nappe (backfold)
V St. Bernard nappe ⎱ Michabel nappe
IV Monte Rosa nappe ⎰
III Monte Leone nappe ⎱ Simplon
II Lebendun nappe ⎰ nappes
I Antigorio nappe

} Pennine nappes

# Chapter VI

# A TRIP ON A VOLCANO

## INTRODUCTION

My chance companion in the gloomy dining room of *Albergo Eremo* was anxious to know why I had been staying at that lonely place for a week already. When he learned I was a stone hunter his immediate reply came: "What minerals are you looking for?" Now this is a question which a geologist is often asked. To make clear, in a few words, the aim of a scientific excursion is by no means an easy task. Moreover, most people expect to hear that the main point concerned has an economic background. Therefore, personally I am accustomed to answer that sort of queries very briefly and always in the same stereotyped way. Thus, in this particular case, too, I simply declared to be interested in: "limestone".

This time, however, my new friend became a bit sarcastic. He emptied his glass of *Lacrima Christi* and opened an attack on me: "Is that your sort of joke trying to make me believe you are looking for limestone on the top of a volcano?"

Actually, however, my short explanation covered part of the truth. Hence, I kindly invited him for a little trip the next day with the full assurance of collecting a greater weight of limestone than he would be able to carry home. For ages ago our old volcano had faithfully thrown up a great many limestone blocks from a depth of between two and three miles. "How do you know the material was derived from that depth" was the next question, and several others followed. So it happened that a harmless tourist — who came only for the purpose of admiring one night the twinkling lights of Naples and the glowing lava flows in the dark crater — joined me for several days and enthusiastically climbed the steep inner wall of the *Atrio del Cavallo* and *Valle dell'Inferno* as well as the deep gullies along the outer slope of *Monte Somma* (fig. 66), thrilled by the gradually accumulating evidence of the complicated and fascinating history of this volcanic pile.

### A FIRST RECONNAISSANCE

It is not so difficult to provide a general idea of the situation around and below Vesuvius. Most tourists to this region will visit the peninsula south of Naples with its beautifully situated places like Amalfi and Sorrento. The whole peninsula is built up of a complex of limestone strata, gently

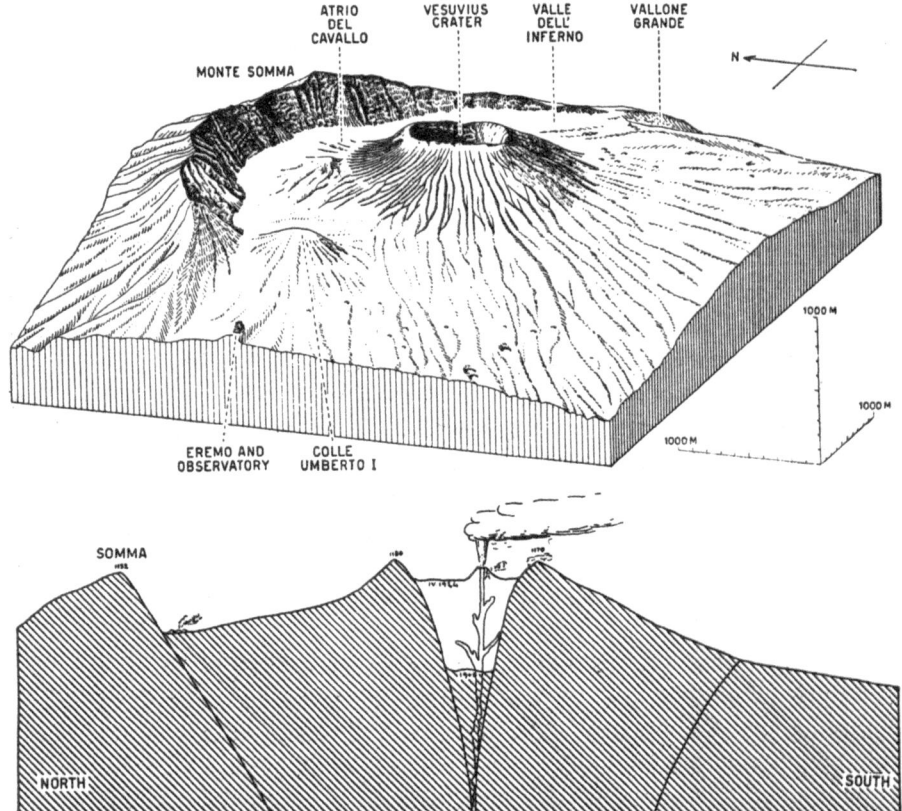

Fig. 66. Block-diagram and section of Monte Somma and Vesuvius; situation in the year 1909. (Drawing by Th. à Th. van der Hoop).

dipping northwestwards, i.e. towards Vesuvius. From a geological mapping of the wider surroundings we can make it plausible firstly that these limestone layers must still be present as far as Naples and secondly that they have not been appreciably influenced by downward faulting or other possible disturbances.

Hence, the depth of these layers below Mount Vesuvius can be easily estimated. Moreover, this led Rittmann, a Swiss petrologist, to locate the depth from which the lavas of Vesuvius are erupted.

The volcano itself is built up of alternating layers of lava — which flowed from a central crater or from an opening in the flank of the mountain — and loose fragments mostly derived from heavy explosive eruptions when much new and old lava is thrown up high in the air: ash, lapilli, volcanic bombs and material scoured off from the walls of the so-called eruption vent. The eruption vent forms the connection between the crater or volcano at the surface and a postulated deep-seated reservoir filled with heated rock material in a fluid or at least in a potentially fluid state. When this material flows out at the surface people call it *lava*, when it is still in its deep-seated reservoir geologists call it *magma* and the loose products are called *pyroclastics*.

Rocks produced by the consolidation of magma are called *igneous rocks*, no matter whether the magma consolidated in the open air or in the earth's crust.

Do not ask at once how big the magma reservoir is, and how it came into

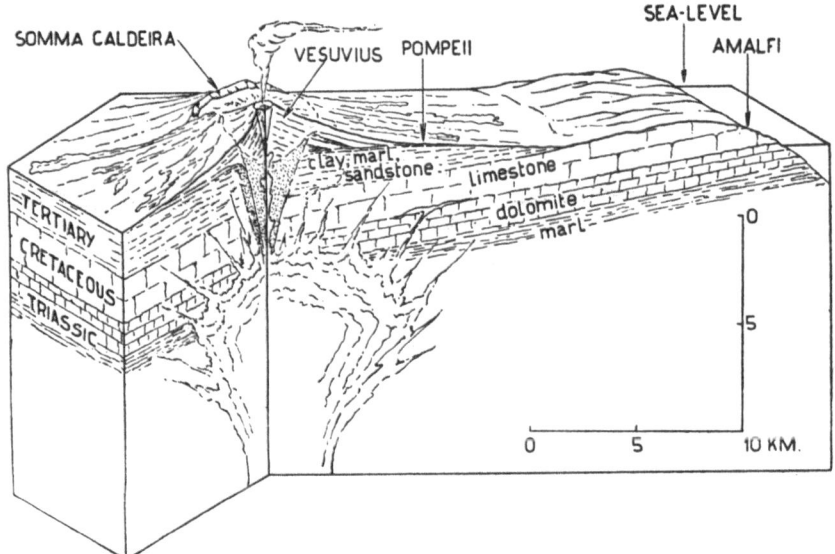

Fig. 67. Schematic block-diagram of the Somma-Vesuvius volcano.

being. These are questions which will be attacked later on. The first problem which will be considered is how deep below Vesuvius is the top of the magma reservoir. During an eruption blocks from Tertiary, Cretaceous and Triassic strata in descending order underlying the volcano, are

Umbgrove                                                        8 ejected (fig. 67). They are found in great quantities among the tuffs of Monte Somma (Pl. VIII, A). Fragments derived from Tertiary strata show little, if any, magmatic influence. Blocks of Cretaceous limestone, too, are mostly unaltered. In exceptional cases they are recrystallized into crystalline limestone. They have been in contact with the magma only during the short time of their ascent in the volcanic vent during an eruption when they were torn off from the walls of the vent.

In strong contrast to the Cretaceous limestone the blocks of Triassic dolomite are thoroughly influenced by chemical processes. Many beautiful silicate minerals are developed in them and not seldom the rock is entirely metamorphosed into silicate rock. Pl. VIII, B, shows an example viz. dark-green crystals of a mineral called vesuvianite. The Triassic dolomite must have been subjected to a close and long-enduring contact with magmatic emanations, hot gases given off by the magma in the reservoir. Hence, it is concluded that the upper part of the magma reservoir has reached the Triassic strata and is gradually assimilating them. As already mentioned, the limestone and dolomite strata crop out at the surface in the peninsula of Sorrento and Amalfi. Their thickness and the angle of their north-westward dip are therefore known. From these data the depth of the top of the magma reservoir below Vesuvius is estimated by Rittmann at approximately 5 kilometres below sea-level (fig. 67).

The first eruptions of the Somma volcano began some 12000 years ago and continued until a cone 2000 metres high had been built up. The present shape of the big crater or caldera of Monte Somma originated during a large destructive eruption of the year A.D. 79. At that time Pompeï was covered by pumice material, while Herculanum was buried under mud-flows. After a long period of quiescence the still active Vesuvius volcano grew up in the centre of the Somma caldera (fig. 66, 67).

From an examination of the outer slopes of the volcano it appears that Monte Somma consists of three successive parts termed primeval, ancient and young Somma respectively. Three eruption phases can be distinguished in the history of young Somma. The last one was the eruption of the year 79. A careful examination of products ejected during the successive outbursts reveal a remarkable phenomenon. Fragments of Triassic dolomite from the deeper and older parts of Monte Somma do not show the strong chemical alteration which is so characteristic of those in the upper and younger parts. Therefore, it is concluded that at the time of the eruptions of primeval and ancient Somma the top of the magma reservoir was deeper than now, i.e. below the level of the Triassic dolomite under Mount Vesuvius. In other words, the top of the magma reservoir has

Plate VIII

(A) Exposure of Somma tuffs near the observatory (Eremo, Vesuvius).

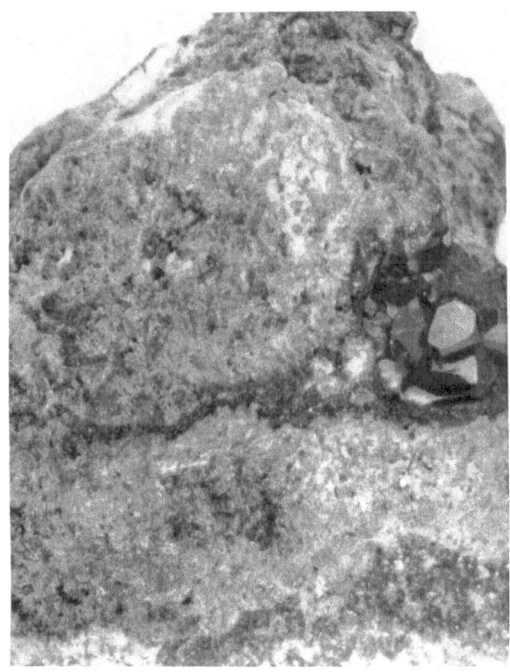

(B) Vesuvianite crystal in Triassic dolomite ejected from about 5000 metres below sea-level (collection Institute of Mines, Delft).

(C) Leucite-lava from Mount Vesuvius (collection Institute of Mines, Delft).

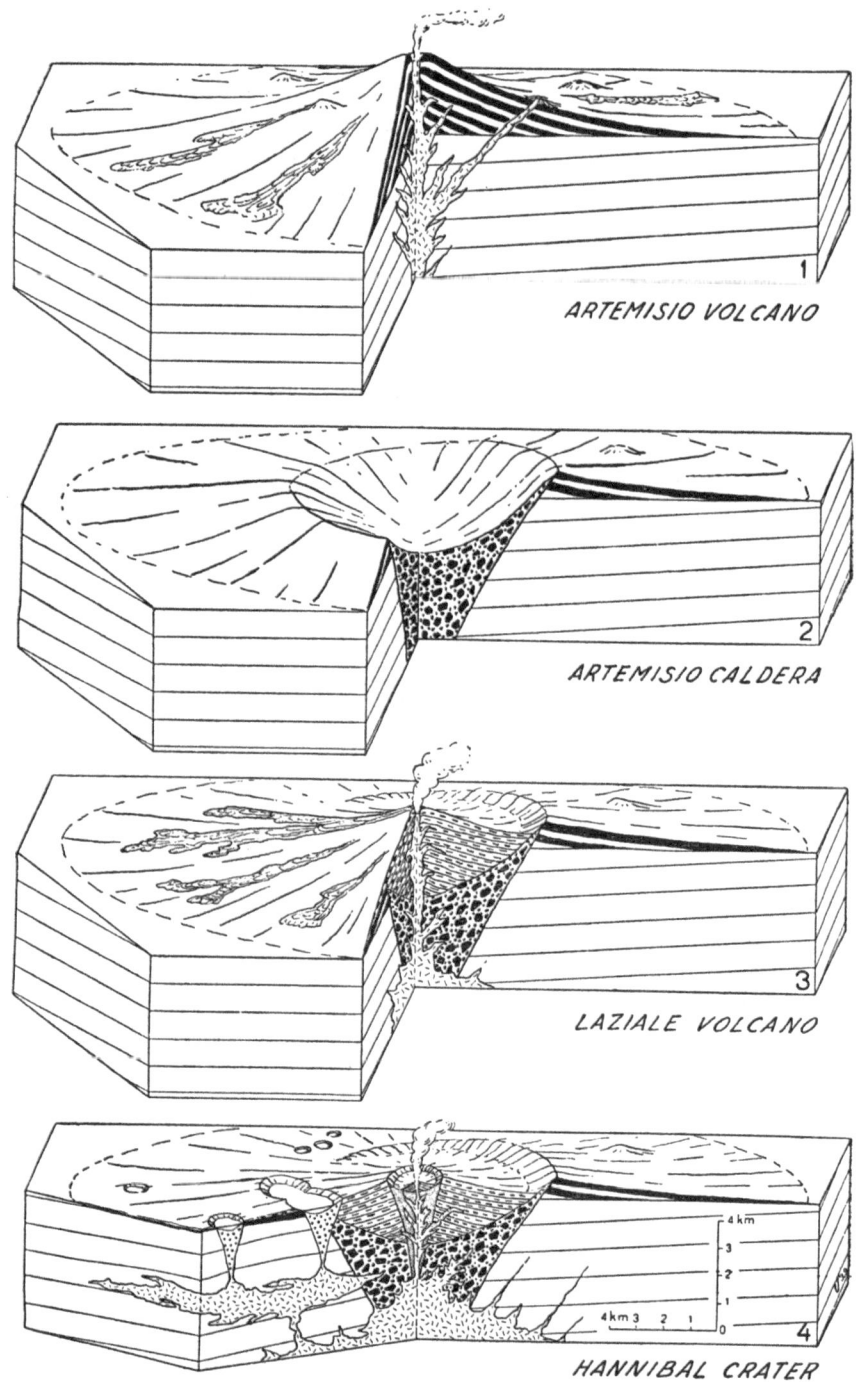

1 ARTEMISIO VOLCANO

2 ARTEMISIO CALDERA

3 LAZIALE VOLCANO

4 HANNIBAL CRATER

Fig. 68. Schematic representation of the history of the Laziale volcano, southeast of Rome, Italy.

gradually forced its way higher up; about 1000 metres during the last 10,000 years.

At a certain stage during a gigantic outburst the top of the Somma volcano was blown off and the great Somma caldera came into being. Similarly, when the colossal explosion of Krakatao took place in August 1883 the greater part of the volcano disappeared and a submarine hollow took its place. Examination of the deposited eruption material showed that only a small percentage of material from the original volcano can be found. So, the greater part of the now missing cone was not blown away but disappeared by collapse, when the big crater formed at the end of the eruption. Again the same conclusion was reached from an examination of Great Crater Lake, Oregon, and its surroundings.

From the inclination of the inner wall of a crater it is generally deduced that the eruption-vent has the shape of an inverted cone. This idea has been expressed in figures 66 and 67. It remains to be seen, however, whether this holds good concerning big craters of the type named calderas. We shall have to return to this point later.

As already mentioned renewed activity in the Somma caldera gradually built up the Vesuvius volcano in its centre.

Numerous volcanoes have one or more younger eruption cones in the centre of an older and much bigger crater. Remnants of old lava streams and cinder layers demonstrate the former existence of an original volcano of considerable height which was decapitated afterwards during a violent explosive outburst.

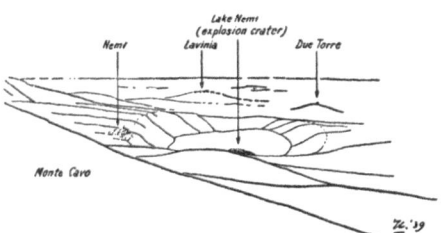

Fig. 69. Lake Nemi, an explosion crater.

A similar sequence of events is displayed — for example — by the large volcano Laziale, southeast of Rome. The series of blocks of fig. 68 represent the original Artemisio volcano (1), the formation of the Artemisio caldera (2), building up of the central Laziale volcano (3), and the origin of the central vent known as Hannibal crater. The volcano is now extinct, but the remnant of the last central eruption cone, the colle Vescovo, is still found in the centre. One of the last events in the history of this volcanic area was the formation of a number of explosion craters. Two of them, the Albano and Nemi craters, are occupied by lakes (fig. 69). The block-diagram, fig. 68 (4) suggests the idea that these lake-craters originated from a separate magma reservoir of comparatively small dimensions.

It is known from several examples that magma often intrudes between sedimentary strata. The overlying part of the earth's crust sometimes becomes arched upwards by the intruding magma of the so-called *laccolith*. According to Rittmann, the island Ischia which is one of the larger islands bordering the south-western side of the gulf of Naples, is underlain by a laccolith.

If the intruding magma takes the shape of plate-like bodies between

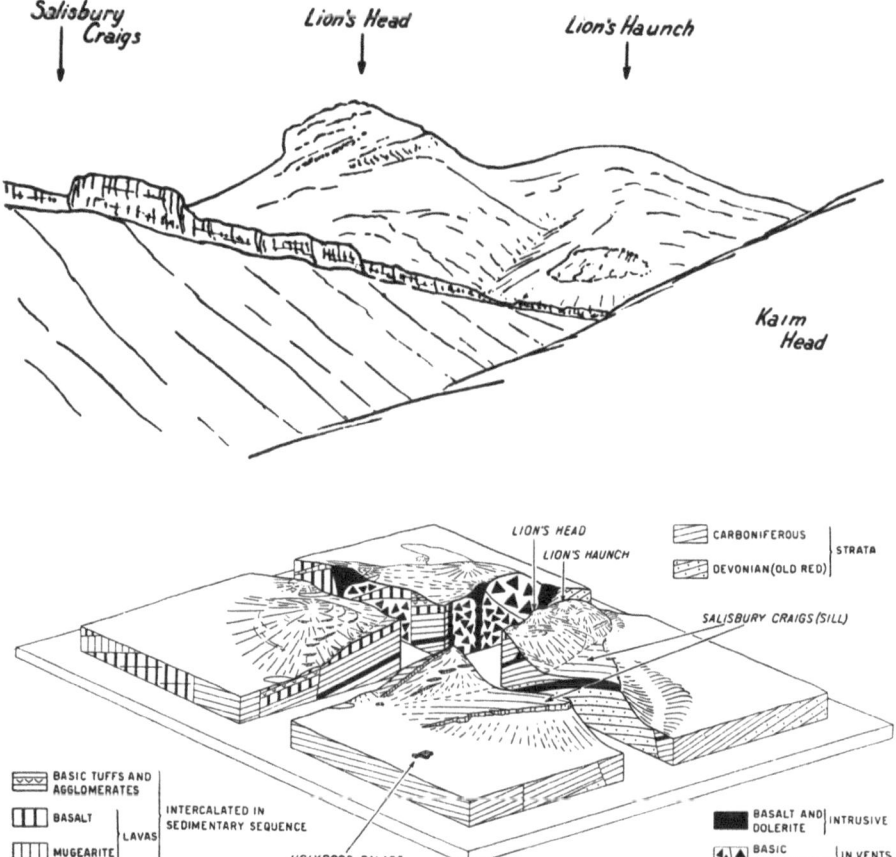

Fig. 70. Salisbury craigs, a sill belonging to the volcanic complex near Edinburgh. The scenery shown in the upper figure is explained by the block-diagram below.

pre-existing strata it is called a *sill*. Fig. 70 portrays a sill belonging to an old volcanic complex in Edinburgh laid bare by erosion, viz. the Carboniferous volcano of Arthur's seat.

When plate-like intrusions of magma cut straight or obliquely across sedimentary strata or volcanic bodies they are called *dykes*. Fig. 71 shows a dyke cutting across the lava flows and tuffs of Monte Somma as well as a few small sills which are offshoots from the dyke.

The dome-shaped hill between the Somma wall and Vesuvius, named colle Umberto I (see fig. 66), represents a lava plug or *tholoid*, a type of formation found with a great number of volcanoes. The material of such a tholoid accumulates where viscous lava appears at the surface. When this type of molten material is thrust up from below great quantities of lava blocks may crumble off from the developing plug. In cases where these can slide down along the flanks of a volcano highly heated gas escapes from the lava and this, together with the finer lava particles which are separated from each other by a cushion of compressed gas, rushes down along the slope of the volcano as an awe-inspiring hot avalanche or *nuée ardente*.

Fig. 71. A dyke and offshoots (sills) in Somma tuffs, Atrio del Cavallo.

Different types of volcanic forms depend mainly on the quantities and character of the lavas and loose products erupted. Gently sloping shield-shaped volcanoes may originate where highly fluid basaltic lavas come to the surface. They are characteristic of the central part of the Pacific. The "girdle of fire" surrounding the Pacific Ocean is, however, characterized by the predominance of so-called calc-alkaline lavas. These are in strong contrast with the basaltic or so-called alkaline lavas of the central Pacific and tend to form steep cones built up of lava flows and pyroclastics. Remarkably enough petrologists call the magma clan of the girdle of fire "Pacific", whereas they have given the name "Atlantic" to the clan of alkali rocks, some of which are found in the central Pacific. Their excuse for doing so is that Atlantic rocks are more widespread and characteristic in the islands of the Atlantic ocean.

Another factor of importance is the percentage of gas in the lava. The higher the gas content the more strongly explosive the eruption. Since the notorious eruption of 1906 the deep crater of Vesuvius has been gradually

Plate IX

*(A)* Strombolian eruption from the central conelet, Vesuvius crater.

*(B)* Vesuvius crater in 1938 with central eruption conelet and lava flows of the ropy type. (After an oil painting by the present author).

filled up by the outpourings of lava and the regular action of one or more small eruption conelets.

Minor eruptions recurring rhythmically at short intervals are characteristic of the Stromboli volcano in the Tyrrhenian Sea. A similar eruption phase in other volcanoes is therefore called a strombolian type of eruption. For many years the conelet in the centre of the Vesuvius crater behaved in that way. Clots of incandescent lava were blown out at regular intervals of a few minutes, to form scoria and volcanic bombs (Plate IX, A). Some times lava flows were also released.

These usually behave like a slowly moving viscous mass and owing to the tranquil escape of the gas the congealed lava surfaces are generally wrinkled into corded and ropy shapes (Plate IX, B). When the gas escapes in sudden and more violent bursts the partly congealed crust of the lava flow breaks into a wild assemblage of scoracious blocks. Mount Etna usually shows this sort of block-lava flows. In 1929 the Vesuvius crater was filled up to such a high level that lava flowed over the lower part of the crater rim descending into the Valle dell' Inferno. When the allied armies entered Italy in 1944 a new violent eruption took place again throwing out the contents of the crater. Apart from this explosive paroxysm Mount Vesuvius showed its effusive action by sending big lava flows down the slopes.

Generally, a continental volcano starts as a cinder cone built up of fragmental material. Monte Nuovo, the most recent volcanic body in the surroundings of Naples, was built up in this way, in September 1538. Similarly Paricutin volcano, in Mexico, sprang up suddenly in the middle of a field in February 1943, almost under the feet of a peasant who was ploughing it. Soon after its birth as a cinder cone, however, a succession of lava flows characterized its further history.

Not far from Monte Nuovo is Solfatara, one of the twelve small volcanic ruins west of Naples, which is well known to tourists. The last traces of volcanic action in this crater have not yet died out and consist of hot and sulphurous exhalations escaping from several small vents in the floor. It is after the name of this crater that volcanoes in either a dormant or nearly extinct phase, showing the same type of fumarolic action, are said to be in a solfataric stage. It is difficult to ascertain whether a quiet volcano is dormant or extinct. The enormous eruption of Krakatao in 1883 took place after a rest of several hundred years. No eruptions were known from Palowe, E. Indies, since historical times, yet unexpectedly a severe eruption started in 1928. And we know that the Somma-Vesuvius volcano had also several long periods of repose, for example from 1139–1631, and at the time of the disastrous outburst of the year 79 the flanks of the mountain as well as the crater were covered by dense vegetation. Other volcanoes display a more regular rhythm in their activity. Mount Hecla in Iceland usually has a great outburst once a century. The list of eruption years of Mount Kelut, Java, shows that a new outburst is due in the near future, for eruptions are known from the following years: 1811 – 1826 – 1835 – 1848 – 1864 – 1901 – 1919. Each volcano has its own peculiarities due to the character of the magma, the nature and thickness of the overlying pile of crustal rocks and several other factors.

When basaltic magma reaches the surface it has a temperature of 1000° C or more. At a depth of 2–5 kilometres the temperature of the country rocks surrounding the top of a magma reservoir can be estimated at 80–170° C. Evidently, the original magma must come from much deeper realms. On the other hand the present situation in which the magma is surrounded by a comparatively cold wall must result in a gradual cooling of the magma. As a result of the chilling convection currents will arise while the liquid magma will tend to crystallize and ultimately solidify. The first crystals to form are of relatively heavy minerals which may sink in the magma.

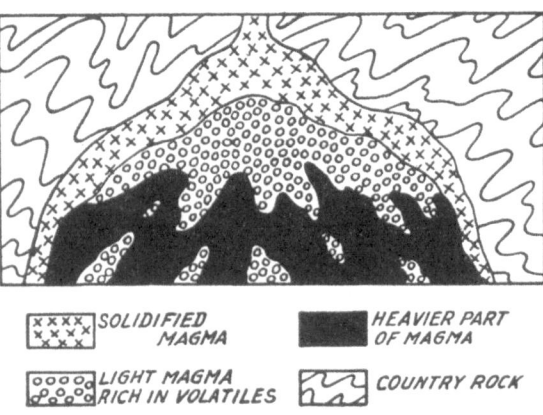

SOLIDIFIED MAGMA

LIGHT MAGMA RICH IN VOLATILES

HEAVIER PART OF MAGMA

COUNTRY ROCK

Fig. 72. Schematic representation of the upper part of a magma reservoir.

Later crystals may be less dense than the magma and will tend to float to the top. The volatile constituents which are dissolved in the liquid magma, like carbonic acid gas in soda-water, will also tend to rise and accumulate in the top of the magma reservoir (fig. 72). This process is termed gravitational or fractional differentiation. As the crystals do not contain dissolved gas, the volatile products are gradually concentrated in the dwindling remains of molten material. Hence, the pressure of the gasses tends to rise. Eventually, when the pressure in the top of a magma reservoir surpasses a certain limit an eruption will start. And in this process it is the fluxing action of the interacting volatile constituents, which probably plays a role of great importance in melting the plug of cooled lavas in the volcanic vent and in clearing the way for the eruption of the deeper parts of the magma reservoir. Only when the pressure of the magma has decreased so far that it cannot lift itself to the crater rim will the eruption end and the vent become plugged with congealed lava.

The residual liquid will again be affected by the same sort of processes. Thus, during the process of differentiation — or fractional crystallization — magmas — and lavas — of contrasted composition may be generated from an originally homogeneous magma.

How deep is a magma reservoir and how did it come into being? Before trying to answer these questions it is appropriate to pay attention first to a few remarkable features in the sector between a volcano at the surface and the top of its magma reservoir at depth.

### REMARKABLE STRUCTURES UNDER VOLCANOES

Suppose we could cut off the Somma-Vesuvius from its basement in order to examine a section below the volcano. Actually, nature has accomplished such an imaginary experiment in several old and deeply eroded volcanic districts. These enable us to examine the structures between a volcano and its underlying magma reservoir.

In the history of Vesuvius one of the striking events was the upward migration of its magma reservoir. Apparently such an upward displacement was in some volcanoes accompanied by the formation of very remar-

Fig. 73. Cone-sheets exposed along the coast east of Faskadale, Ardnamurchan.

kable cone-shaped cracks which were injected by magma (fig. 73). These
features have been described from Nigeria, Ireland and western Scotland.
On a geological map the solidified injections in the cracks appear as
sets of discontinuous concentric outcrops of lava-sheets. Each set is
circular or elliptical in outline (fig. 74). The lava-sheets all dip towards a

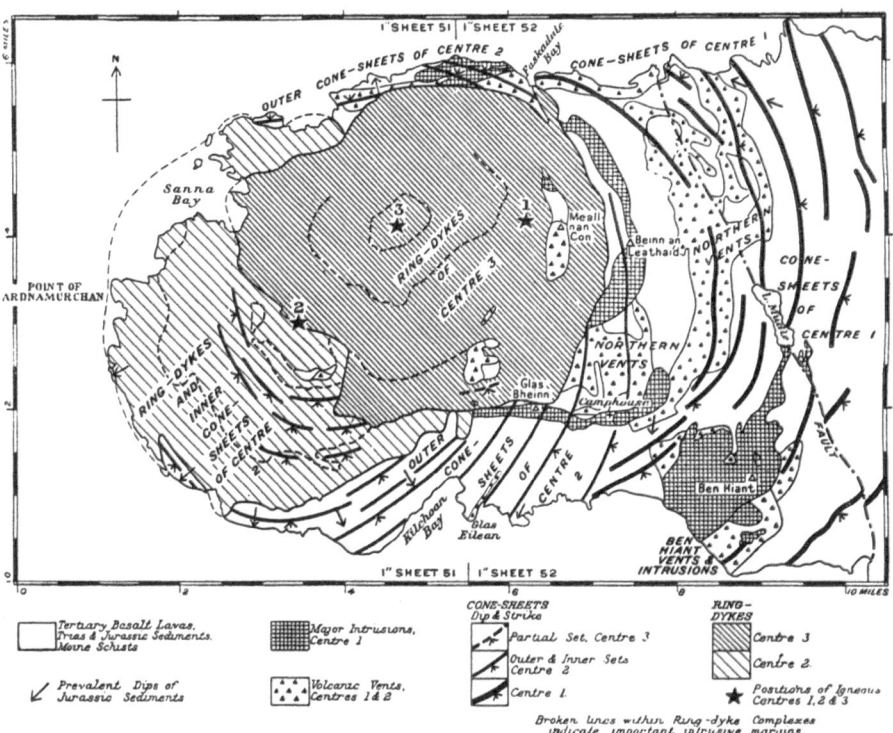

Fig. 74. Map of ring complexes of Ardnamurchan, Scotland. (After J. E. Richey).

common centre, but the central area at the surface is free of these so-called
cone-sheets (fig. 74). These cone-shaped lava injections cut across the
previously existing country rocks. The latter show not the slightest sign
of being shattered by this process and the regular shape and position of
the crack is not influenced by the material or structure of the intruded
rock, whether it consists of friable sediments or compact older intrusive
rocks. Nor is the course of the cone-shaped crack deviated by pre-
existing lines of weakness in the country rocks. So, the suggestion that
the cone-sheets were emplaced along the steeply dipping bedding planes
of material which tumbled down in the deep cone-shaped vent of a
caldera is out of the question. On the contrary, they appear to be upfillings

of real cone-shaped fractures which radiated from a deep-seated centre due to a heavy and sharp "blow" against the top of the magma reservoir. One sometimes sees a conical hole pinched in the window pane of a bus or auto-car. In this case the culprit was a naughty boy who threw a stone at the vehicle. The cone-sheet cracks may also have been caused by fracturing of a quite sudden character due to a strong explosive increase of pressure at the top of the magma reservoir. The cause of the sudden rise of magmatic pressure remains an open question. Some authors consider a process of retrogressive boiling due to progressive cooling of the surroundings to be the main process involved.

We need not think, however, of the emplacement of a set of cone-sheets as one single act. On the contrary, the process of cone-sheet formation and injection was of an intermittent nature, separate sharp blows or bumps each yielding only a small fraction of the total amount of cracks and injected material.

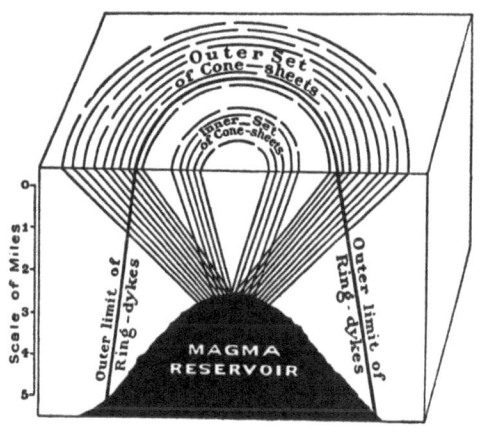

The apex of the inverted cone of cone-sheets would approximately coincide with the top — or with a top — of the magma-reservoir at the time of their origin. The depth could be deduced from the inclination of the cone-shaped cracks. It appears to be of the order of three miles below present sea-level (fig. 75).

Fig. 75. Diagram showing two sets of cone-sheets in relation to inferred magma reservoir. (After J. E. Richey).

Are there cone-sheets below Monte Somma and Vesuvius?

Possibly cone-shaped intrusions formed in advance of some of the more violent eruptions. Tentatively a few have been drawn in fig. 67. It remains a remarkable fact, however, that cone-sheets are known only from a few old and deeply denuded volcanic districts. Apparently they originate only when special conditions of shearing, strength of the crustal material, shape and depth of the top of the magma reservoir are fulfilled. Kuenen demonstrated that as a consequence of cone-sheet injection the upper wall of the crack is lifted vertically roughly to the thickness of the sheet injected, measured vertically. Hence, the central part within a set of cone-sheets must have been gradually lifted thousands of feet, in the course of the long enduring process of cone-sheet injection.

Now let us examine what will happen if the reverse, viz. a decreasing pressure takes place in a magma reservoir. Volcanic outbursts and crystallization of material may cause a decrease of pressure in the magma reservoir As a result part of its roof may collapse.

In some cases decrease of magmatic pressure within the reservoir may cause a more or less circular system of nearly vertical cracks above the top of a magma reservoir.

Intrusion of magma along such cracks would give origin to circular or elliptical dykes which have been described from several districts as *ring-dykes*.

The cylindrical block within the circular cracks could sink down into the magma reservoir. As a result an "empty" room would originate either somewhere in the crust if the crack did not reach the surface, or if it carried right through a depression would form at the surface. In the first case the "empty room" would be filled up by magma intruding from below concomitant with the sinking of the central plug. In Bailey's opinion the Glencoe mountains in Scotland show the result of such a process (fig. 77, A). In the second case when the whole roof collapses a circular volcano-tectonic sink or *cauldron* would originate at the surface.

The Ossipee Mountains in the east central part of New Hampshire, offer an example which is illustrated by the map and a series of schematical sections of fig. 76.

According to Kingsley a thick series of volcanic rocks formed on a primeval surface of granite which is probably of Pre-Cambrian age (block 1). A magma reservoir is postulated somewhere deep below. For some unknown reason — let us suppose decrease in the magmatic pressure — the roof of the magma reservoir collapsed along a cylindrical crack, which reached the surface of that time. The cylindrical mass within the ring-shaped crack was then down-faulted. Following or possibly accompanying the down-faulting of the cauldron magmatic material was intruded along the peripheral cracks, forming the present ring-dyke (block 2). Then, a granitic magma intruded the middle of the subsided area (block 3). Probably all these events occurred in Devonian times, about 320 million years ago. Block 4 portrays the present situation, after considerable erosion has removed large quantities of the softer granites, leaving part of the down-faulted more resistant volcanics as higher parts of the scenery.

According to Anderson's mathematical deduction of the fracture systems above a magma reservoir the tension fractures resulting when the magmatic pressure was less than the pressure exerted by the surrounding

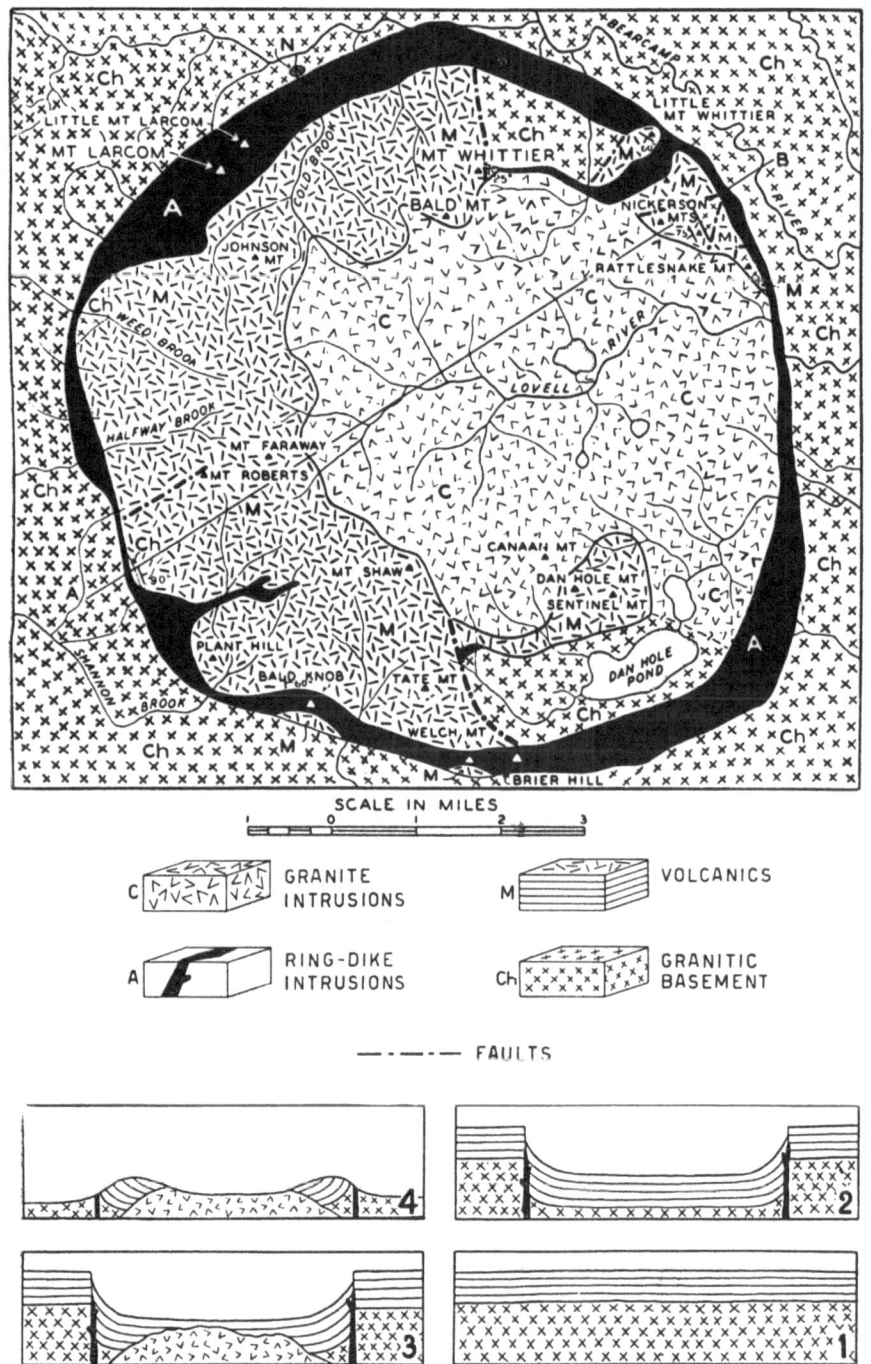

Fig. 76. Geological map of the Ossipee Mountains, New Hampshire. The black notation represents a ring-dyke. The geological history of the region is portrayed by the sections 1–4. (After L. Kingsley).

rocks would be paraboloid convex upward. The theory is not entirely satisfactory; for, in New Hampshire the contacts of ring-dykes are not dipping outward but are essentially vertical. In Billings' opinion the origin of ring-dykes might in some cases result as extension fractures due to

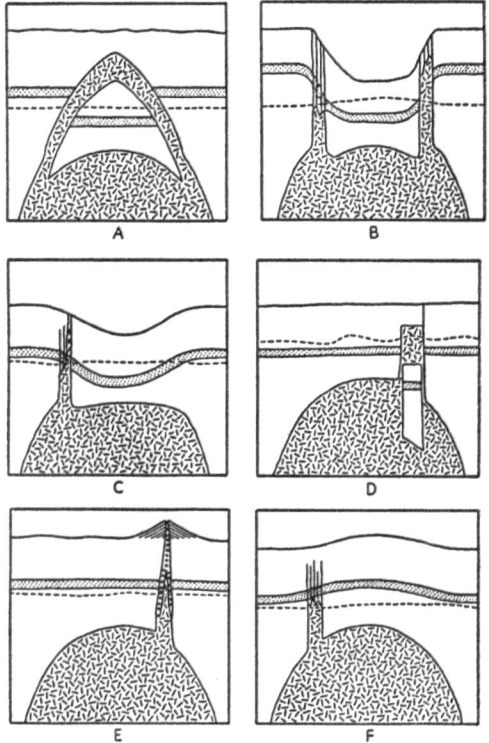

increasing instead of reduced pressure in the magma reservoir. Originally the roof might even be domed up (fig. 77, F). Still, however, after the magma has established a passage to the surface the roof rocks will subside into the magma-reservoir below and the final condition will be as illustrated by fig. 77, B. The ring-fractures will not necessarily encompass the possible 360 degrees of a complete circle (fig. 77 C, E, and F) in which case no cauldron subsidence will occur.

The original ring-shaped fractures were widened by magmatic action. The main process is thought to have been a gradual shattering of the original rocks by thermal action, penetration of gases and magma-tongues into the walls of country rock, sinking and finally melting and assimilation of the blocks in the

Fig. 77. Different types of cauldrons and ring-dykes. Broken line is present erosion surface. (After M. P. Billings).

ascending and fluid magma. This process was called *piece-meal stoping* by Daly.

It is difficult to ascertain whether the process of ring-dyke intrusion accompanied by the subsidence of a central body has taken place in one of our recent or sub-recent volcanic districts. A huge crater the horizontal diameter of which largely surpasses the vertical dimensions of the crater walls is termed a *caldera*. As already mentioned its origin is due to a gigantic volcanic outburst. Several calderas have formed in historical times and in all these cases the accompanying eruptions discharged vast quantities of ash and other material of magmatic origin. This shows that the reservoir lost a large volume of magma prior to the collapse of the cone. Surely some material, perhaps even the greater part of the material which once took the place of the cone-shaped vent, may have tumbled down into the reservoir. Possibly, however, some calderas owe their origin to the subsidence *en masse* of a cylindrical body enclosed by vertical or even outward dipping fractures. In such a case the terms caldera and cauldron would be synonyms. As to Monte Somma it seems not warranted to substitute the term caldera by the term cauldron.

### THE LOWER AND UPPER BOUNDARIES OF VOLCANOES

*The root of a volcano.*

What is the shape of the magma reservoir feeding a volcano, and how did it become emplaced? What is the relation between volcanism and the enormous massifs of plutonic rocks that cover large areas on the continents? How did they become emplaced? These tantalizing questions are not so easy to answer. As a matter of fact, they are fundamental problems. As in nearly all fundamental problems we must frankly admit that an attempt at an answer only leads to the expression of fundamental uncertainty. Ask an astronomer how the solar system — including the earth and the moon — came into being. Probably he will review a great number of rivalling theories that have been advanced in the course of time but finally he will conclude that none of the existing theories is satisfactory, though many notable advances have been made in the last twenty years. Ask a biologist questions like these: What is life? How did it originate? What is the mechanism of evolution? What is the fundamental difference between a living and a non-living system? Perhaps he will tell you a long story but his final conclusion will be to admit that fundamental uncertainties are at the root of all your questions. Ask a geologist how the interior of the earth is constituted. He will be able to explain the great contributions made by seismic, magnetic and gravity researches, he will advance very stimulating physical and chemical considerations. In conclusion he will unite all results into a synthetic picture, but he will have to admit several fundamental uncertainties inherent in his "model".

Still, if we want to understand the phenomena of volcanism and plutonism, we must start from a certain mental picture of the earth's interior.

As a matter of fact, different models have been proposed and our knowledge is inadequate for the attainment of a definite conclusion. Though

we may add this to the list of subjects on which uncertainty prevails there
is no need for complete agnosticism. On the contrary, a great number of
geological and geophysical data have furnished converging evidence
suggesting that a shell of light granitic composition overlies a dense shell
of olivine-basaltic composition. Further, it is believed that the upper
granite layer is almost entirely restricted to the continents. The light and
silica-rich granitic rocks are usually indicated by the name *sial* and it is a
custom to speak of the basic rocks as *sima*. The sialic layer is absent
beneath the Pacific basin proper. As regards the Atlantic and the Indian
Oceans, their bottoms are thought to consist of thin layers of sial overlying
material of basaltic composition. Perhaps the basaltic material under a
continent is built up of two shells differing in chemical composition, viz. a

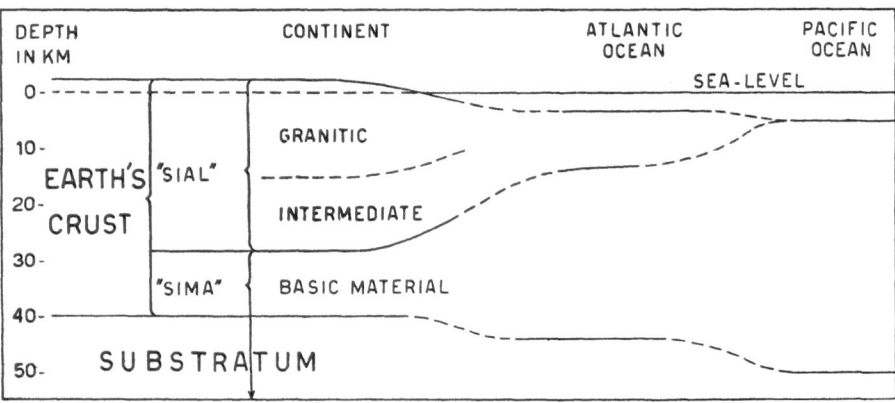

Fig. 78. Crustal layers in a continent and in two oceanic areas.

lower one of olivine-basalt and an upper one called tholeitic as we shall see
later on. Moreover, there probably exists a still deeper shell of material
which is even denser and still more basic than basalt, the so-called perido-
titic layer. But we are not concerned with it in the following considerations.
Fig. 78 summarizes these views schematically (see also fig. 12). The upper
portion of these shells is in a consolidated crystalline state. This part is
termed the earth's crust and it is supposed to rest on material in a viscous
state termed the *substratum*. Geophysical evidence shows the crust to be
roughly in isostatic balance. The continents float high because the sialic
rocks are light, the oceans are deep because their floors are underlain
mainly or entirely by heavy material.

Going into the tantalizing problem of the origin and actual distribution
of the different earth-shells would lead us to the distant and turbulent
period of our planet's infancy. There is, however, no need to discuss this

fundamental problem now [1]). Suffice it to say that personally I am inclined to believe that a thin and continuous sial layer has been thickened into the discontinuous continental masses by a process of buckling and drifting in early Pre-Cambrian times.

For the moment we are only interested in the question whether the model "works" for an understanding of volcanic and plutonic phenomena.

Beyond doubt the plutonic and volcanic rocks accessible to observation at the surface belong to two main types. Plutonic intrusive bodies are dominantly of granitic composition, whereas the dominant extrusive volcanic rock is of basaltic composition. According to Barth about 95 percent by volume of the intrusive rocks are granitic, and about 98 percent of the extrusive rocks are basaltic. Rocks of intermediate chemical composition are comparatively rare.

A further striking fact is that basaltic rocks have been found on oceanic islands as well as on the continents, where in all geological periods they penetrated the sialic rocks and their sedimentary cover in vast quantities.

Conversely, the oldest and deepest visible parts of the continents show a striking dominance of granitic rocks, which in turn are absent from true oceanic islands. Let us now suppose that the layer of basaltic material of the substratum is tapped by a deep reaching fissure or rectilinear crack leading up to the surface. An intrusion of magma rising up from abyssal depth was termed an *abyssolith* by Daly (fig. 79). Possibly the mag-

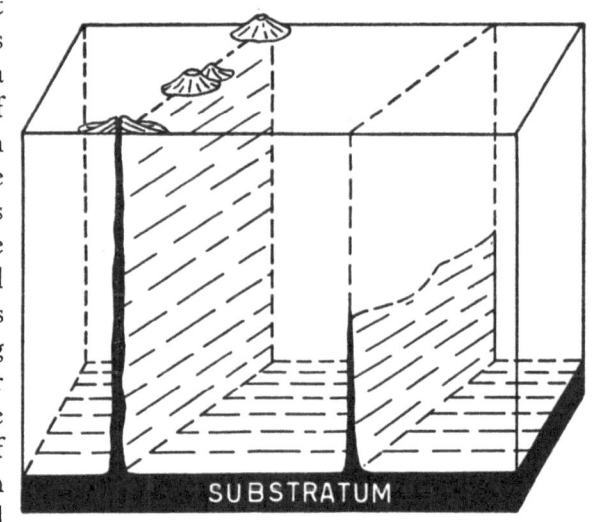

Fig. 79. The root of a volcano, a so-called abyssolith.

matic intrusion becomes wider and bigger by magmatic stoping, the same process which we learned from the widening of ring-dykes. By contamination and assimilation of crustal rocks the crack may become widened

---

[1] For a detailed discussion of this question see: The Pulse of the Earth 1947, pp. 241–250, and Journal of Geology, vol. 244, 1946, pp. 169–178.

so as to form a magma pocket or magma reservoir. For some time Daly propagated the thesis that the substratum would consist of basaltic material in a non-crystalline state. Due to the high pressure at that depth the magma would be very viscous, more or less like glass — hence Daly's term "glassy substratum" — but it would be in a potentially eruptible state. By the load of the overlying crust it would immediately be driven to the surface and form outpourings of lava if it were tapped by a fissure dissecting the crust. For several reasons the once widely favoured idea of a world-circling layer of glassy basaltic material has been given up, even by Daly. However, circumstances must be such that adequate supplies of basaltic magma must be available from great depth and practically everywhere, locally and temporarily. How and why are questions which we must add to our list of fundamental uncertainties.

But even admitting these uncertain aspects, the crust-substratum hypothesis clears up several remarkable phenomena, while the distribution of sialic and simatic shells is a good working hypothesis for the explanation of several volcanic and plutonic phenomena, as we shall see in the following sections.

*An upper boundary of volcanic growth.*

Volcanoes cannot indefinitely grow aloft. They are bound to certain limits of height, and it appears as if the crust-substratum hypothesis almost automatically accounts for the maximum height found in continental and oceanic volcanoes.

Suppose a volcano is built on a deep fissure dissecting the continental crust completely, the magma coming up from a depth of about 40 kilo-

Fig. 80. Mount Etna, as seen *en route* from Catania to Casa Cantoniera.

metres (fig. 79). According to prevailing opinions the upper part of the granitic crust has a density of 2.7. The lower 25 kilometres are supposed to consist of heavier material with density 3. The total pressure exerted by the crust on the underlying magma, which is thought to have a basaltic composition with density 2.7 in the molten condition, would be $15 \times 2.7 + 25 \times 3 = 115.5$. The maximum height, x, to which magma could ascend when forced upward through the fissure can easily be estimated. For we have $115.5 = 2.7 (x + 40)$; whence $x = 2.78$ kilometres. Volatile constituents in the magma may lower its density and help its ascent to a level a few hundreds metres higher up.

One of the highest basaltic volcanoes in Europe is Mount Etna reaching an elevation of about 3.3 kilometres (fig. 80). Probably it has reached the maximum elevation attainable by a basaltic cone on a continent. As a result the ascending lavas tried to find another way out causing flank eruptions. Probably irregular sill-shaped offshoots from the liquid column

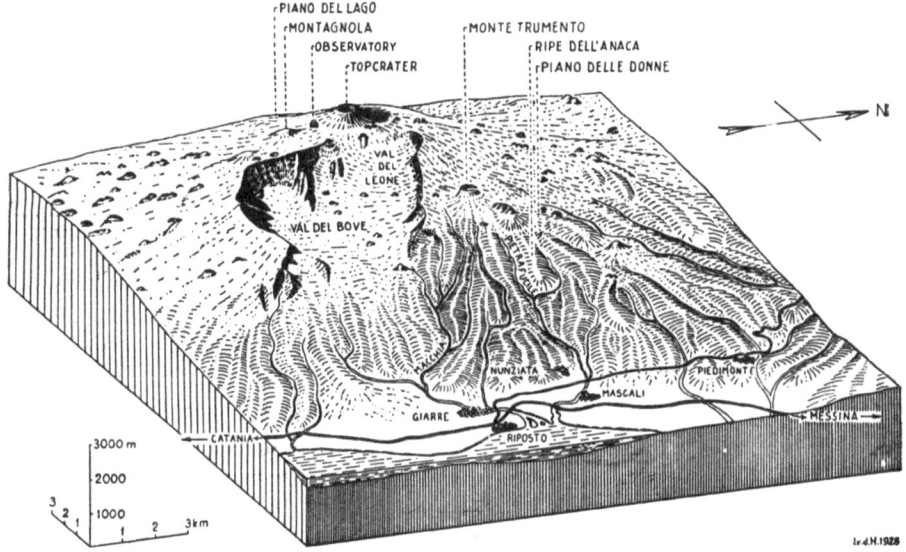

Fig. 81. Block-diagram of Mount Etna. (Drawing by Th. à Th. van der Hoop).

in the vent found a way out towards the flanks of the volcano. Characteristic, indeed, for Mount Etna are numerous radial fissures that have opened on its slopes. Along the eastern side of the mountain even a great block bounded by radial fissures collapsed; the resulting wound is the famous Val del Bove (fig. 81).

Fissure-eruptions on the flank of a volcano display a curious phenomenon (see Plate X, A). A series of craters open along the fissure during an eruption. The uppermost craters are just deep vents originated from mere explosive action. Proceeding downwards the next crater rims are built up of small explosive products called lapilli. Then follow craters surrounded by bigger ejectamenta like volcanic bombs. Finally, lava flows were produced from the lowermost crater. The explanation of this striking arrangement of craters displaying an eruptive sequence which runs from explosive to effusive is shown in fig. 82.

A similar phenomenon was seen during the 1926 eruption of the Batur volcano in the island Bali.

Fig. 82. Arrangement of different types of craters characteristic for a fissure eruption along the flank of a volcano.

The surface of a lava-flow cools and solidifies rather rapidly to form a hard stony crust. If this crust breaks at the snout of the flow and the still molten inner part continues further down the slope the cooled crust

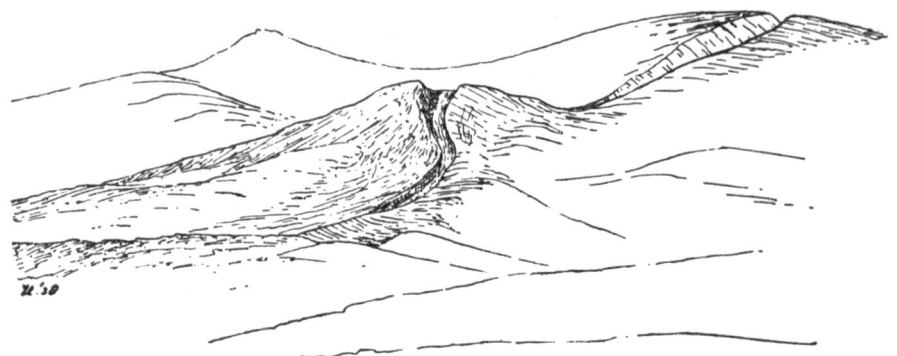

Fig. 83. Lower part of the fissure eruption of 1910 showing lava gully. Mount Etna near Casa Cantoniera.

is emptied of its contents. The solid carapace remains as a hollow tube. Sometimes the roof of the tube collapses while the lateral walls remain as boundaries of a huge gully of solid lava. Fig. 83 and Plate X, B shows an example from Mount Etna; the lava originated from the lowermost crater belonging to the fissure eruption of the year 1910.

So, a basaltic volcano like Mount Etna was possibly built up on a deep reaching fissure by so-called abyssolithic injection.

Plate X

*(A)* Linear arrangement of craters belonging to the fissure eruption of the year 1910, Mount Etna.

*(B)* Lava gully which originated by the collapse of the roof of a lava tunnel, Mount Etna, near Casa Cantoniera. The lava flowed out from the lowermost crater of the fissure eruption of the year 1910. See figs. 82 and 83.

The elevation of the volcanoes of the central Pacific — when measured from the summit to their base on the sea-floor — greatly surpasses the average elevation of continental volcanoes. An amount of 6 to 8 kilometres is not an exception.

If these basaltic volcanoes have been built up by abyssolithic injection through fissures dissecting a sub-Pacific crust — which consists of basaltic material with density 3 — it would follow that a crust with thickness x exerts a pressure of $3x + 5$ on the eruptable basalt under the crust. If we take the elevation of the volcano at 8 kilometres and the depth of the ocean at 5 kilometres, we have $3x + 5 = 2.7 (x + 8)$; whence $x = 55$ kilometres.

If this deduction is correct it would follow that the solid crust under the Pacific is thicker than in continental areas.

### PONTIFFS AND SOAKS

It was suggested that magma is occasionally injected along fissures from a deep layer of basaltic composition. Gradually the material in the reservoir will cool and more and more minerals will crystallize (see p. 120). Due to the sinking down of heavy minerals and the upward movement of the lighter ones the magma will gradually differentiate. In this way a great diversity of "consanguinous" rocks evolves from an original uniform parental magma. The final products vary from dark and heavy, comparatively silica-poor so-called basic rocks to lighter so-called "acid" rocks containing a higher percentage of silica. Some petrologists even hold that all volcanic and plutonic rocks derive from one parental magma of olivine-basaltic composition. To be sure, a few different rock clans can be recognized but even these different clans are considered by them as the result of processes of secondary importance. Becke who introduced the terms Atlantic and Pacific magma-clans chose the names because — as already mentioned — examples of the first clan of consanguinous rocks are known from certain islands in the Atlantic Ocean, whereas calc-alkaline rocks of the Pacific clan characterize most of the volcanoes on the island-arcs which surround the Pacific Ocean. Generally, however, the volcanoes within the Pacific Basin proper are composed of Atlantic lavas. It is generally accepted that the atlantic clan of rocks evolved from an alkaline parental magma with the composition of an olivine-basalt. It is generally agreed that the crust below the Pacific Basin consists of basaltic material. So, even if large quantities of crustal material be assimilated by an abyssolith no other rocks than those of the Atlantic clan can originate. But what if the magma reservoir is emplaced in a continental area where the upper part of the terrestrial crust consists of acid silica-rich material of granitic composition?

In that case the chemical and mineralogical composition of the original basic parental magma would become acidified by assimilation of parts of the pre-existing acid rocks and as a result the magma would produce a Pacific clan of rocks. Is granite also a differentiation product from basaltic magma? Some petrologists have, indeed, defended this thesis. Others hold that occasionally and to a very limited extent this may hold true but certainly not for the majority of the granite massifs. We have arrived at another topic belonging to the category of fundamental uncertainties.

Moreover, the question how a granite massif became emplaced and what happened to the pre-existing country rocks is at present one of the most controversial and keenly debated problems of petrology. Petrologists are divided in a few camps that attack each others opinions. Mostly — I am glad to say — the fight is a friendly though earnest game frequently mixed with jokes and irony, but occasionally the writings tend to become sarcastic and passionate; expressions of feeling are used in lieu of scientific arguments, reminding one of political discussions in time of war rather than of a search for truth.

The controversy is as old as the science of plutonic rocks. More than a hundred and fifty years ago Hutton started to teach the igneous origin of granite. According to this great founder of geology granite intruded the crustal layers in a liquid state, as magma. Consolidation of granitic magma would convert it into a crystalline plutonic body — more or less in the same way as fluid lava when cooling becomes a solid rock. This is now called the dogmatic or pontificating point of view of the magmatists or *pontiffs*. The most extreme opinion on the other side is that there does not exist a single granite or diorite in the whole earth's crust which ever was "igneous", i.e. a molten mass. In this view granite is no igneous rock but a metamorphic rock which has always been formed by the gradual transformation of pre-existing rocks due to the soaking action of hot solutions or emanations which rose into the invaded crustal parts from deeper realms. Those holding this opinion belong to the group which Bowen called *soaks*. Some modern authorities even hold that the transformation took place by ionic transfer in the solid state.

A process by which pre-existing solid country rocks are gradually converted into granite without passing through a magmatic stage is termed *granitization*. In the opinion of Wegmann progressive metamorphism of the country rocks (termed *migmatization*) proceeds through the country rocks like waves or fronts ascending from deep realms towards higher levels of the earth's crust. As the migmatite front advances granitization proceeds, ultimately leaving granitic rocks behind. Granite bodies generally appear to become wider when proceeding towards greater depth and frequently they have enormous dimensions. They are usually supposed to rise up from deep parts of the earth's crust; hence their name *batholiths*. The question as to what happened to the original country rocks that once occupied the space now taken up by a granite batholith is a serious difficulty for the magmatist. The doctrine of granitization solves the problem of room automatically.

The doctrine of fronts is strongly adhered to by Backlund and it is especially Reynolds who has given a great deal of detailed evidence from observation in the field as well as under the microscope. Nobody ever witnessed the formation of a granitic rock. From both camps *pro* and *contra* arguments based on a limited number of facts are advanced in such an abundance of papers and books that the non-petrologist sometimes feels like the *academicus* in fig. 89. What will be the impression on an unbiased outsider when reading that the basic front of migmatists is called by Bowen "a basic affront to the intelligence of the geologic fraternity", and Reynolds' answer: "This is accusing Nature of an intentional breach of politeness"? Will Read's consoling suggestion that probably there are granites *and* granites clear up much of the controversy for him? Still, however, it should be possible to reconcile the strong contrast expressed by transformists and magmatists. Let us take just a few concrete examples.

In the very old pre-Cambrian areas of Sweden and Finland the country rocks are, as it were, soaked by granitic material. In many places the country rocks have become altered entirely into granite (Fig. 84). In a more advanced stage of granitization than the example shown in Fig. 84 the rock still represents vague remnants or "ghosts" of the structural pattern of the original schists. Finally, even these remnants vanish and 'the rock has entirely become a granite. I wonder if careful inspection in the field would not convince even a stubborn magmatist that in the case under consideration the granite has become emplaced by transformation of the original country rock into granite. A process of molecular change from one material into the other seems to have taken place. We may here leave the question open whether the transformation took place by the action of hot solutions or emanations, by ionic transfer in a dry state or not. It is no use for magmatists to attempt throwing doubt on the actual occurrence of fronts. They have been described and portrayed from several areas In a recent paper, for example, Holmes and Reynolds have given what seems convincing evidence of the existence of fronts in the Dalradian of Co. Donegal.

In other areas, on the other hand, careful examination in the field gives

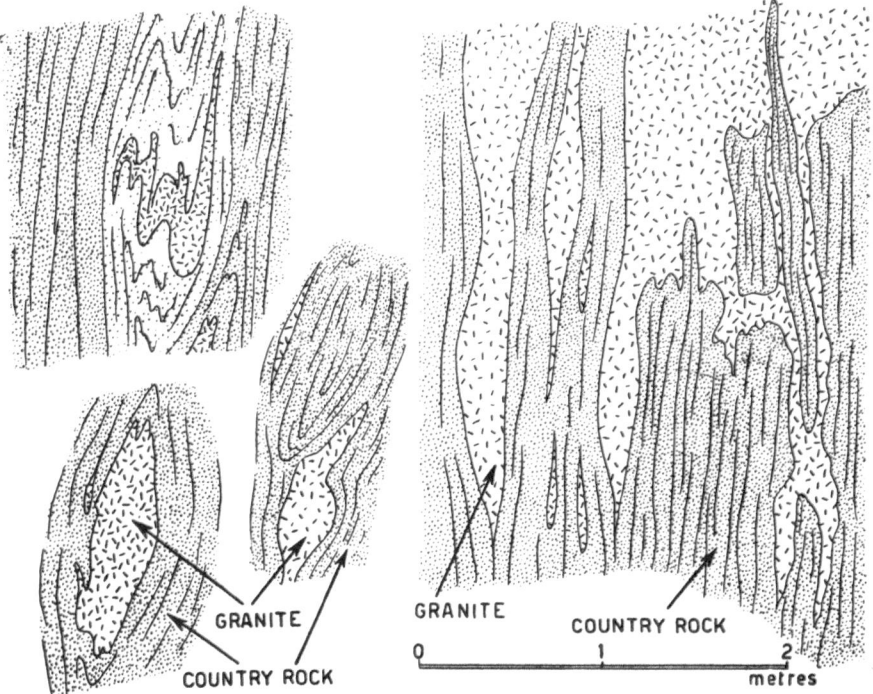

GRANITE

GRANITE     COUNTRY ROCK

COUNTRY ROCK

0     1     2

metres

Fig. 84. Granitization of very old (pre-Cambrian) schists near Masthuggskyrkan, Göteborg, south-western Sweden.

strong support to a process of magmatic stoping which means that one cannot possibly escape the conclusion that at least the granite in the immediate surroundings of the roof of some of the Cornish granites was fluid or viscous and therefore in a magmatic state. Moreover, in several places the tectonic setting of the granite as well as of its surroundings cannot be explained without accepting not only that it was fluid or viscous at a certain stage but also that it exerted an active pressure on the surrounding rocks. As an example fig. 85 shows a block-diagram of a granite mass in the southeastern Cevennes, France. To all appearance the roof of the granite mass became arched and parts of it were lifted up, whereas the structure of the granite is adapted to the surrounding frame work. Let us now leave out of consideration whether contamination and assimi-

lation of country rock has taken place or not in this case. What matters is
that the present situation leads to the acceptance of active movement
of this part of the granite body at a certain stage. This can hardly be
denied even if a "soak" would say that still the primary source of this
magma was migma and that granitization took place at a lower level.
Therefore, I personally cannot refrain from having the strong conviction

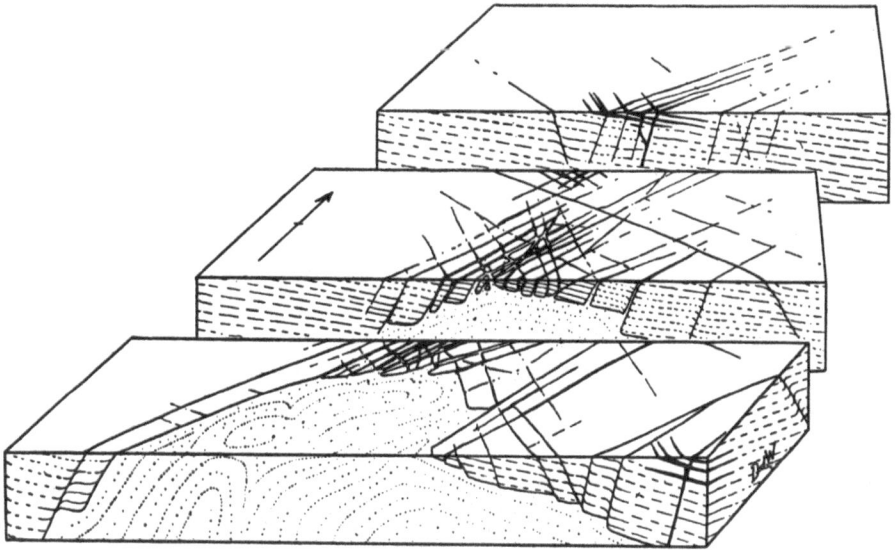

Fig. 85. Block-diagram of the northern part of the granite massif of Mount Aigoual, Cevennes,
France. (After D. de Waard).

that there are, indeed, granites *and* granites.

Possibly the different types occur at certain depths of the crust and at
certain stages of the evolution of the area in which they occur and so their
origin is in some way interrelated.

Very probably accumulation of further detailed evidence collected
in the field as well as under the microscope will bridge the contrasting and
extreme opinions expressed by the terms: pontiffs and soaks.

### VOLCANISM AND PLUTONISM

Fig. 89 strongly suggests that granite is primarily a product of granitization: *per migma ad magma*. Another fundamental point expressed in the
same illustration is the contrast between Vulcanus and Pluto, between
volcanism and plutonism.

It is a striking fact that the great granite batholiths always occur in the fold-belts. Moreover, their shape is frequently elongated along the axes of these belts and the time of their emplacement is closely associated with epochs of folding and mountain-building. All this demonstrates a genetic relation between the formation of fold-belts and granite massifs. Apparently the granite bodies originated and gradually expanded towards the surface when the crust was thickened by a process of down-buckling to form a mountain root. It should be held in mind that a fold-belt is always born out of a geosyncline (see p. 13) which means that the time of folding is always preceded by a long period of subsidence of that part of the crust. The subsiding furrow was filled up with sediments and, remarkably enough, volcanic products as well as outpourings and intrusion of basic to ultrabasic rocks are found mingling with the sediments of the geosyncline especially in its axial part. Acid plutonic rocks, however, never penetrate the geosyncline at this stage of its development. Conversely, they make their appearence during and after its folding. And in these periods outpouring of lava does not occur in these areas. Fig. 9 portrays this phenomenon in a diagrammatic manner. Block I represents a geosyncline with accompanying volcanism and intrusion of basic magma. The contents of the geosyncline, including the magmatic products, are folded in the stage represented by block II. It shows the local thickening of the crust penetrating into realms of progressively higher temperature. Doubtless the root does not remain intact under such conditions. What exactly happens is difficult to say. To avoid points of controversy let us define the processes taking place in the root by the rather vague term disintegration. Evidently, however, disintegration of the root involves at least partial upward migration of sialic material. Due to these thermal processes the deeper parts of the folded contents of the geosyncline become strongly metamorphosed. At higher levels the metamorphism is generally less intense. But, locally uprising sialic material causes the ultimate emplacement of "batholithic" granite bodies, especially in the axial region of the fold belt.

**THERMAL SURFACES**

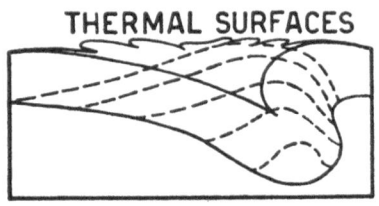

Fig. 86. Thermal surfaces rising up from a mountain-root.

Fig. 86 illustrates the uprising isothermal surfaces emanating from a mountain-root. Proceeding upward and outward the degree of regional metamorphism will decrease. Due to the arched shape of the isothermal surfaces several more or less parallel zones of equal thermal influence may be expected at the surface. Actually, supporting evidence is found in the more deeply eroded mountain belts. It has been known for several decades, that certain critical, newly-formed minerals are indicative of different

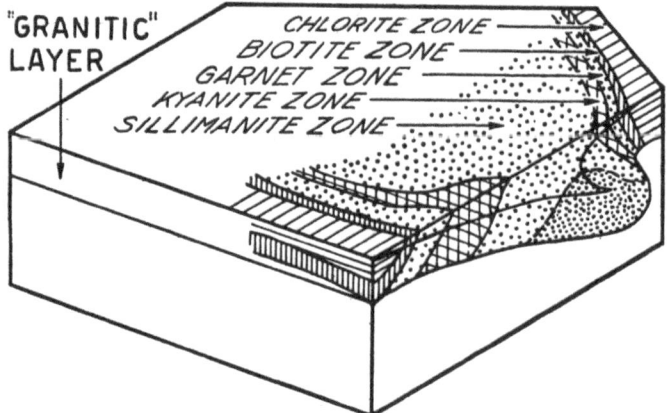

Fig. 87. Thermal zones in a fold-belt. (Based on A. Q. Kennedy).

degrees of metamorphism in rocks which originally had the composition of normal argillaceous sediments. In the Caledonides of Scotland several zones of progressive metamorphism have been investigated and mapped by Barrow, Elles, Read, Reynolds, Kennedy and others. In a very diagrammatic way the principle of indication of the zones, each of which has been named after a special index mineral — as well as their interpretation as thermal zones — is shown in fig. 87 which is based on two recent papers of Kennedy's. Once more a few thermal surfaces are represented by the broken lines

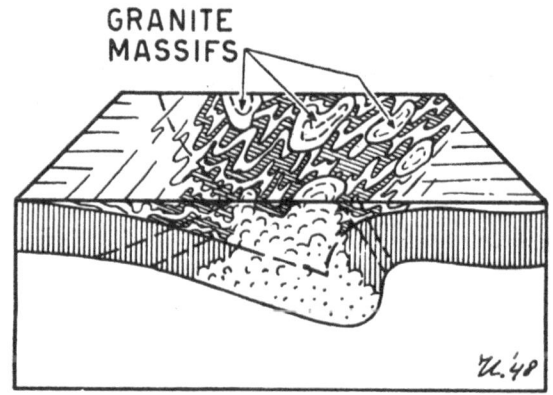

Fig. 88. Eroded fold-belt, showing granite-batholiths and mountain-root.

in fig. 88. However, the main purpose of this diagram is to show the genetic relation between the formation of a mountain root and the emplacement of granite massifs in a fold-belt (Compare also fig. 9 block II).

The place and time relation between granite massifs and fold-belts may be regarded as firmly established facts. In respect of both space and time, volcanic phenomena show quite different relations. Volcanic phenomena are found either in the geosynclinal stage of a fold-belt or long after the time of root formation and folding — when crustal movements of a quite different nature accompanied by updoming and faulting take place — or outside the fold-belts mainly in areas where again updoming of the crust accompanied by faulting and possibly by rifting, prevail. It appears that the emplacement of a granite massif is an extremely slow and long-enduring process accompanied by "waves" of thermal action involving migmatization of the country rocks. Moreover, the ascent of granite batholiths is fundamentally different from the origin of the magma reservoir of a volcano. By some hydrostatic, thermal or other unknown control the ascent of a granite batholith is arrested before it reaches the surface of the crust.

In short, there appear to be two fundamentally different groups of manifestations of activity from deep realms involving intrusion, extrusion and associated phenomena. Granite batholiths owe their origin to processes resulting from the regional and temporary disintegration of the granitic or sialic upper layer of the earth's crust where it is thickened in the downward bulge of a mountain root. Conversely, volcanic phenomena result from local (or regional) and temporary remelting of a basaltic earth shell and they are generally associated with the formation of deep-reaching cracks or faults.

Geological field evidence therefore strongly supports the classification presented by Kennedy and Anderson, who recognise two groups of "igneous" phenomena called the volcanic and plutonic associations respectively.

To the volcanic phenomena belong the true volcanic vents, dykes, cone-sheets, ring-dykes, laccoliths and sills. Basic, basaltic magma is the primary source of the volcanic association of rocks. The comparatively small amount of acid volcanic rocks may conveniently be explained partly by crystallization-differentiation of basaltic magma, partly by assimilation of acid country rocks in the melt of the magma reservoir.

To the plutonic association belong the granitic and granodioritic rocks of batholiths in fold-belts. They developed from processes accompanying the disintegration of the so-called "granitic" layer in areas where the latter was thickened by a process of down-buckling so as to form a mountain-root. The contrast between the two groups is clearly depicted in fig. 89.

Whether the granitic batholiths were emplaced by emanations, soaking, or ionic transfer and whether they were partly or entirely viscous or

fluid, solid or volatile during these processes are questions which undoubtedly will be cleared up by further detailed investigations in the
field and under the microscope.

In the meantime all this fits in very well with the model of the earth's

Fig. 89. Illustration displaying Read's classification of three classes of rocks, viz. Neptunic
(sediments), volcanic and plutonic. (Drawing by Dr. Gilbert Wilson, from H. H. Read).

crust as depicted in a previous section, so far as its most salient features
are concerned.

There is still enough room for diversity of opinion as soon as one
enters into details. Kennedy thinks the granitic layer of a continent is
underlain by two types of basaltic layers, the lower layer being of true

olivine-basaltic composition. Magma tapped from that layer could give rise to the Atlantic clan of rocks. Another layer intermediate between the granitic and olivine-basaltic layer, both in respect of place and chemical composition would furnish the parental magma of the Pacific rock clan. In the opinion of others there is no need for the latter assumption, inasmuch as the Pacific rock suites might be explained by assimilation of sialic material by a magma that originally had the composition of olivine-basalt. In the same way they try to explain the origin of rocks belonging to the so-called Mediterranean clan. This, at last, brings us back to the volcano which was our starting point.

### BACK TO MOUNT VESUVIUS

Emerging from the dark realms of Pluto and Vulcanus, we shall pay a final visit to the lavas of Monte Somma and Vesuvius.

Probably the volcano originated on a deep reaching fissure. For reasons which, if set forth *in extenso*, would take us too far away from our present subject it is found that in Pliocene times, some 10 million years ago, the distribution of land and sea was entirely different from what it is now in the surroundings of the Italian volcanoes (see fig. 90). A land area in the present Tyrrhenian sea supplied sediments which were deposited in a sea which extended over parts of what is now Italy. Then, however,the Thyrrenian block started to subside, whereas the sea withdrew from Italy, due to an upward movement of that area. At the boundary between the two blocks with opposed movements deep reaching cracks, or faults, originated. Along these faults the magma forced its way upward, starting to build up volcanic bodies at the surface. The abyssolith became wider and gradually formed a magma reservoir which forced its way higher up. During this process material from pre-existing crustal rocks was assimilated. As a further consequence the chemical composition of the magma changed in the course of time. This process manifested itself also in the chemical and mineralogical composition of the lavas erupted during the long though intermittent history of the volcano. An ever increasing quantity of dolomite material became assimilated by the magmatic melt (see p. 114 and fig. 67). Assimilation of dolomite and limestone is considered by Rittmann to have caused the hydrothermal escape of soda. Consequently, the potash contents of the residual melt became comparatively enriched. Leucite is a typical potash mineral. It is abundant in the lavas as beautiful white crystals regularly bounded by 24 lozenge-shaped faces (Plate VIII, C). Studying the whole suite of lavas of Monte

Somma and Vesuvius in chronological order it appears indeed that they become ever more rich in leucite crystals.

As a matter of fact, this is only part of a chemical and mineralogical story which is much more complicated. Suffice it to mention, finally, that similar potash-rich lavas are known from several other districts, but that they became first known from Italian volcanoes. Hence, P. Niggli introduced the name of Mediterranean rock suites to indicate this sort of lavas, even though they occur far away, for example in southern Celebes, and though not all petrologists agree that the above given explanation of their origin holds for occurrences in other places.

It was a piece of leucitic lava together with a specimen of ejected Cretaceous limestone which my companion took with him when he left *Albergo Eremo.* He wanted to keep them as a pleasant memento of his visit to Mount Vesuvius. For, though his original aspirations were fulfilled — he had enjoyed seeing the twinkling lights of Naples and the glowing lava in the crater — he was still more impressed by the absorbing *Symphony of the Earth.*

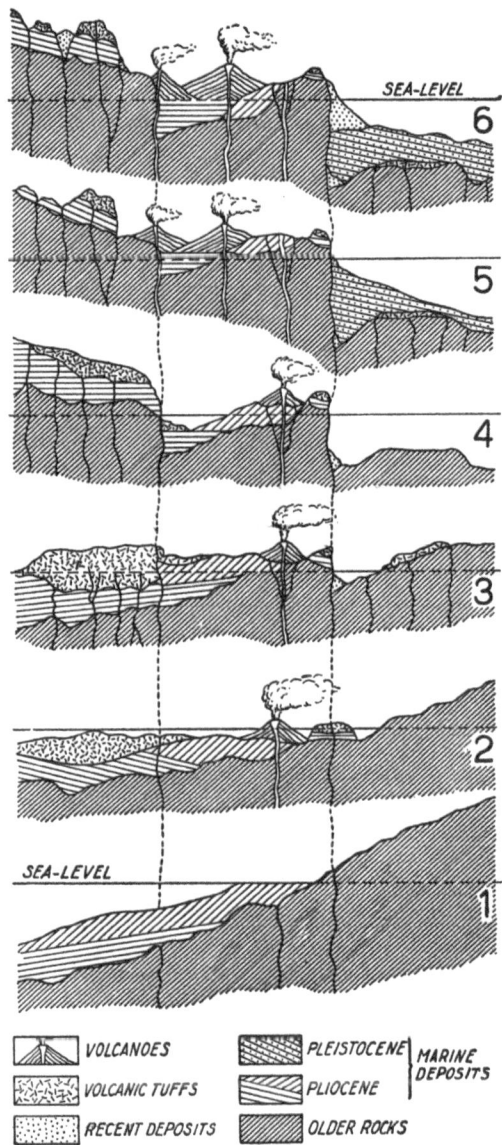

Fig. 90. Evolution of volcanic action along the present Tyrrhenian sea since Pliocene times. (After A. Portis, adapted from A. Sieberg).

## REFERENCES

Though a great number of publications were used in preparing this chapter, only the principal papers mentioned in the text are enumerated in the following list. They may serve as a key for further references.

ALFANO, G. B. und FRIEDLÄNDER, I., *Die Geschichte des Vesuv.* (Berlin 1929).

ANDERSON, E. M., *The dynamics of the formation of cone-sheets, ring-dykes, and cauldron-subsidences.* (R. Soc. Edinburgh 1936, vol. 56, pp. 128–157).

BAILEY, E. B., e.a. *Tertiary and post-Tertiary Geology of Mull, Loch Aline and Oban.* (Mem. Geol. Surv. Scotland 1924).

BAIN, A. D. N., *The younger intrusive rocks of the Kudaru Hills, Nigeria.* (Quart. Journ. Geol. Soc. London 90, 1934).

BILLINGS, P. M., *Mechanics of igneous intrusion in New Hampshire.* (Americ. Journ. Sci. 243A, 1945, p. 40–68).

BOWEN, N. L., *Magmas.* (Bull. Geol. Soc. America 58, 1947, pp. 263–280).

COTTON, C. A., *Volcanoes as landscape forms.* (Whitcombe and Tombs Lim 1944).

DALY, R. A., *Igneous rocks and the depth of the earth* (1933).

DALY, R. A., *The roots of volcanoes.* (Transact. Americ. Geophysical Union, 1938, pp. 35–39).

DALY, R. A., *Volcanism and petrogenesis as illustrated in the Hawaiian islands.* (Bull-Geolog. Soc. America 55, 1944, pp. 1363–1400).

DALY, R. A., *Granite and metasomatism* (Americ. Journal of Science, 247, 1949, pp. 753–778).

GLANGEAUD, L., *Les états de la matiére dans la pétrogenese profonde.* (Experientia III, 1947).

HOLMES, A., *Natural history of granite.* (Nature 155, 1945, p. 312).

HOLMES, A. and REYNOLDS, D. L., *A front of metasomatic metamorphism in the Dalradian of Co. Donegal.* (C. R. Soc. geolog. de Finlande XX, 1947, pp. 26–64).

JAGGAR, T. A., *Origin and development of craters.* (Geol. Soc. of America, Mem. 21, 1947).

KENNEDY, W. Q. and ANDERSON, E. M., *Crustal layers and the origin of magmas.* (Bull. volcanologique, ser. II, T. III 1938, pp. 23–83).

KENNEDY, W. Q., *Crustal layers and the origin of ore deposits.* (Schweiz. Petr. Mitth. 28, 1948, pp. 1–8).

KENNEDY, W. Q., *On the significance of thermal structure in the Scottish Highlands.* (Geolog. Magazine 85, 1948, pp. 229–234; 86, 1949, pp. 43–56).

KINGSLEY, L., *Cauldron subsidence of the Ossipee Mountains.* (Americ. Journ. Sci. 22, 1931, pp. 139–168).

KUENEN, PH. H., *Intrusions of Cone-sheets.* (Geolog. Magazine 74, 1937, pp. 177–183).

MISCH, P., *Metasomatic granitization of batholithic dimensions.* (Americ. Journal of Science, 247, 1949, pp. 209–246, 372–407, and 673–706.

*Origin of Granite.* (Geolog. Soc. of America, Mem. 28, 1948).

PERRIN, R. and ROUBAULT, M., *On the granite problem* (Journ. of Geology, 57, 1949, pp. 357–380).

RAGUIN, E., *Géologie du granite.* (Masson et Cie, Paris 1947).

READ, H. H., *Meditations on granite.* (Proceed. Geol. Association LIV, 1943, LV, 1944).

READ, H. H., *A commentary on place in plutonism.* (Q. Journ. Geol. Soc. London CIV, part 1, 1948).

READ. H. H., *Time in Plutonism* (Abstr. Proceed. Geologic. Soc. of London no. 1450, 1949).

REYNOLDS D. L., *Observations concerning granite*. (Geologie en Mijnbouw, 11, 1949, pp. 241–263)

RITCHEY, J. E. a.o., *The Geology of Ardnamurchan, North-west Mull and Coll*. (Mem. Geol. Surv. Scotland 1930).

RITTMANN, A., *Die geologisch bedingte Evolution und Differentiation des Somma-Vesuvmagmas*. (Zeitsch. f. Vulkanologie XV, 1933, pp. 8–95).

RITTMANN, A., *Vulkane und ihre Tätigkeit*. (Enke, Stuttgart, 1936).

SIEBERG, A., *Einführung in die Erdbeben und Vulkankunde Süditaliens* (1914).

UMBGROVE, J. H. F., *Vulkanische verschijnselen in de omgeving van Napels*. (Tijdschr. Koninkl. Aardrijksk. Genootschap LVI, 1938, pp. 98–105; see also LVII, 1940, pp. 75–80).

WAARD, D. DE, *Tectonics of the Mt. Aigoual pluton in the southeastern Cevennes France*. (Proceed. Kon. Acad. v. Wetensch. Amsterdam LII, 1949, pp. 389, 539–550, and 1027–1038).

WEGMANN, C. E., *Zur Deutung der Migmatite*. (Geol. Rundschau 26, 1935).

WILLIAMS, H., *Crater Lake. The story of its origin*. (Univ. of California Press, 1941).

WILLIAMS, H., *The Geology of Crater Lake National Park, Oregon*. (Carnegie Inst. Washington, Publ. 540, 1942).

# Chapter VII

## LIFE AND ITS EVOLUTION

### INTRODUCTION

There is an old story about an undergraduate and a professor. The former was asked by the latter to give definitions of life and matter. After having ransacked his brains the undergraduate replied that he could not hit upon the answers just then, although — he asserted — he knew them exactly. Which, according to his malicious examiner, was a very great pity as his answers might have been one of the greatest contributions to science.

As a matter of fact, "definitions" of life and matter would mean the solution of two fundamental problems. Later on, when dealing with attempts at formulating living systems as contrasted with non-living systems, we shall learn some of the difficulties involved.

It needs hardly be said that the problems which we shall have to discuss are intimately interwoven with the fundamental problems of human life and destiny in general, with one's attitude concerning social and religious doctrines in particular.

Among the biological attempts at interpretation of life three trends of thought have had a special influence on the philosophical outlook of civilized people during the last century. One is the materialistic point of view. Diametrically opposed to it is the group of vitalistic speculations. A third doctrine, called holism, contains a negation of both materialism and vitalism.

All three were born as purely scientific disciplines and most of their creators and proponents never intended or even dreamt of becoming world reformers or propagators of a revolutionary antagonism. As an exception, General Smuts, the creator of the biological concept of holism, applied the same principle also in South Africa and even in world politics. Generally, when the achievements of science slowly but surely penetrate into wider circles of society and in part become common property, the real meaning of the achievement is either not understood or not properly appreciated. People like to turn an interesting theory into an accepted fact and dogmatize accordingly. A simile is taken for a representation of reality, or a scientific concept is transformed into a propagandistic slogan which only partly covers its original meaning.

Everybody knows how fiercely debated materialism has become in some social, religious and political circles, how the philosophy of Thomas Aquinas is the only one tolerated in other quarters. On several occasions Darwin's theory on evolution was made responsible for irreligion and the glorification of war. Others, overstating and exaggerating some critical arguments, set themselves to denounce the concept of evolution entirely.

We shall not discuss these aspects of the doctrines concerning the relation between life and matter, for it would take us too far away from the domain of science.

Apart from the three theories mentioned so far — or more precisely we should speak of three groups of theories — a new way of looking at the fundamental problems of life and matter is gradually evolving in recent times. If one would have it characterized by a single word I should like to term it the theory of complementarity. It will be discussed in the section "living and non-living systems". Nobody can foresee whether these modern speculations will ever be susceptible to condensation into a social or religious discipline. Remarkably enough, however, they remind one of certain aspects of the Brahmanese Upanishads. Probably most people will find it extremely difficult to abstract all anthropomorphic ideas from scientific concepts of life and matter — all images or pictures which are familiar to us from our daily surroundings. Moreover, it will be asked what beacon is left to man to guide him on his way, and how the results can be united into a harmonious synthesis. An attempt at answering these questions will be given in a final section.

Let us first, however, deal with a few remarkable features as displayed by the evolution of the organisms.

SOME ASPECTS OF EVOLUTION

Imagine the apparatus portrayed in fig. 91 be placed in a slightly inclined position, the aperture upwards. Little pellets are introduced through the opening (A).

In their way downward they meet a great number of obstacles by which they are deviated from their straight course. Whether they turn to the left or to the right is a matter of pure chance. Evidently, however, there is a greater chance of finally arriving in the central part of the box

than at the extreme right or left. An accumulation of a great number of
pellets therefore shows the well-known
appearance shown in fig. 91. The surface
of the accumulation corresponds with a
curve of probability which is the expression of Newton's binomial $(a + b)^n$.

Living beings if examined and measured
in great numbers reveal a similar frequency distribution in the variation of their
characters. Fig. 92 is a frequency curve
obtained by weighing a great number
of beans. Usually such a community is
termed a *population* by biologists.

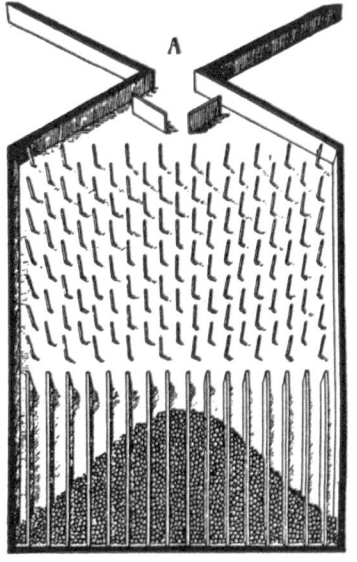

Fig. 91. Frequency distribution demonstrated by Galton's apparatus. (Adapted
from E. Baur)

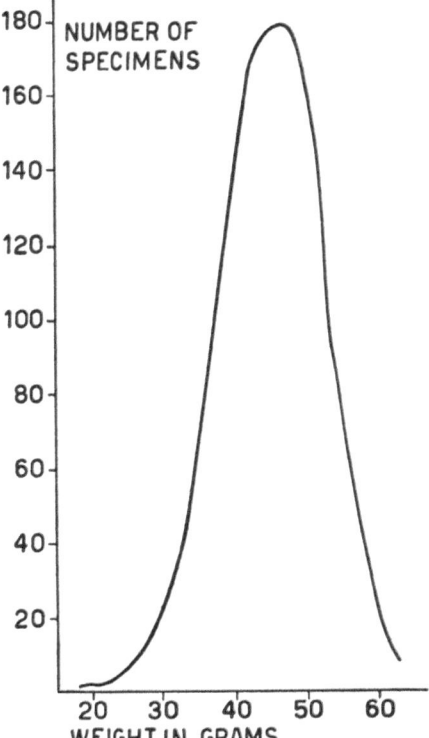

Fig. 92. Frequency curve resulting from
weighing 712 beans. (Adapted after E. Baur).

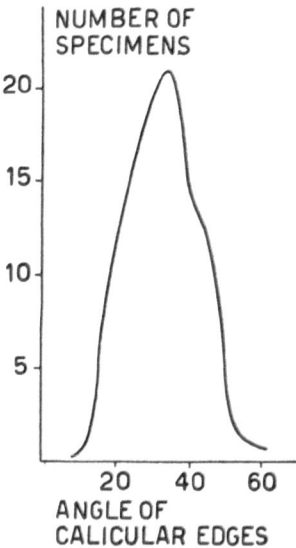

Fig. 93. Frequency curve resulting from
measuring the calicular angle of 144 specimens
of Flabellum altum, a fossil solitary coral from
the Lower Pleistocene Putjangan beds, Java.

Quite similar results are obtained when studying a population of fossils. Flabellum altum is a coral occurring abundantly in the Tertiary and Pleistocene deposits of Java. The calicular angle of the corallum varies between 10° and 60°. This character was plotted against the number of individuals measured as shown in the accompanying graph (fig. 93). Again the same type of frequency distribution is found.

In the next examples we shall consider a sequence of several populations which was carefully collected from a succession of strata covering a considerable span of geological time. This enables us to study the evolutional tendencies of the group inasmuch as these are exhibited in their progressive characters.

In some cases the paleontological record enables us to follow the evolution of a population at close quarters and to ascertain a gradual differentiation of the population into two or more separate crests. Such a crest corresponds to a community which is classified by the taxonomist as a subspecies. The next illustration, fig. 94, portrays the origin of three subspecies

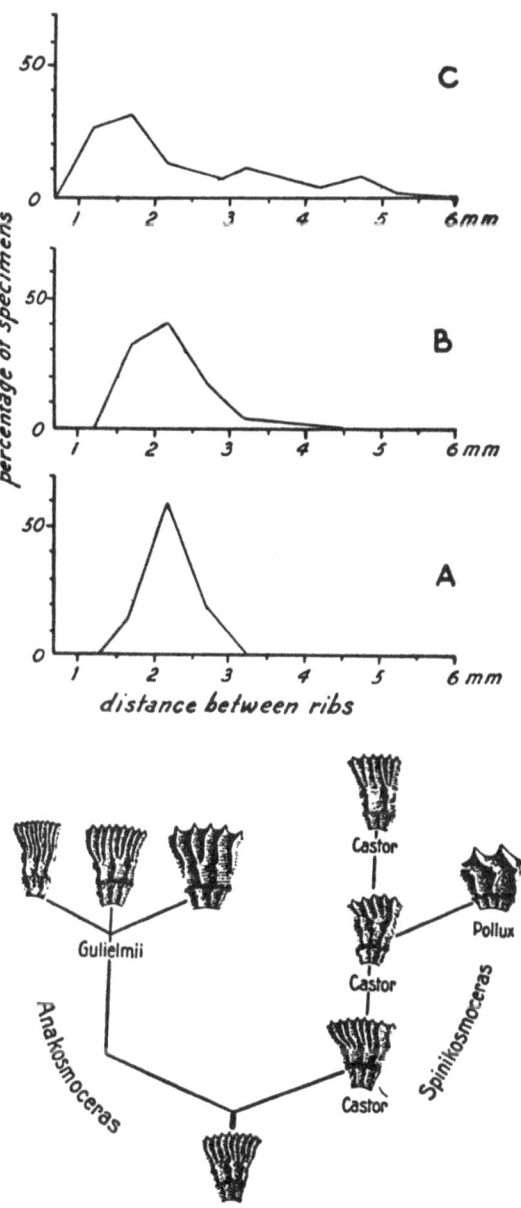

Fig. 94. Evolution of Anakosmoceras gulielmii showing differentiation of the ancestral stock into new subspecies. (Adapted after Brinkmann).

evolving from the ancestral stock of the ammonite Anakosmoceras gulielmii. The material was collected in the Oxford Clay of England and studied in great detail by Brinkmann. A characteristic feature is the presence of radially arranged ribs on the last convolution of the shell. Brinkmann found this character to be constant during a time interval of about 150,000 years represented by the lower 6 metres of exposed strata. At about 7 metres above the base of the stratigraphic section, however, the distances between the ribs exhibit a variation ranging from $1/2$ to 6 millimetres, whereas the original variation ranges only from

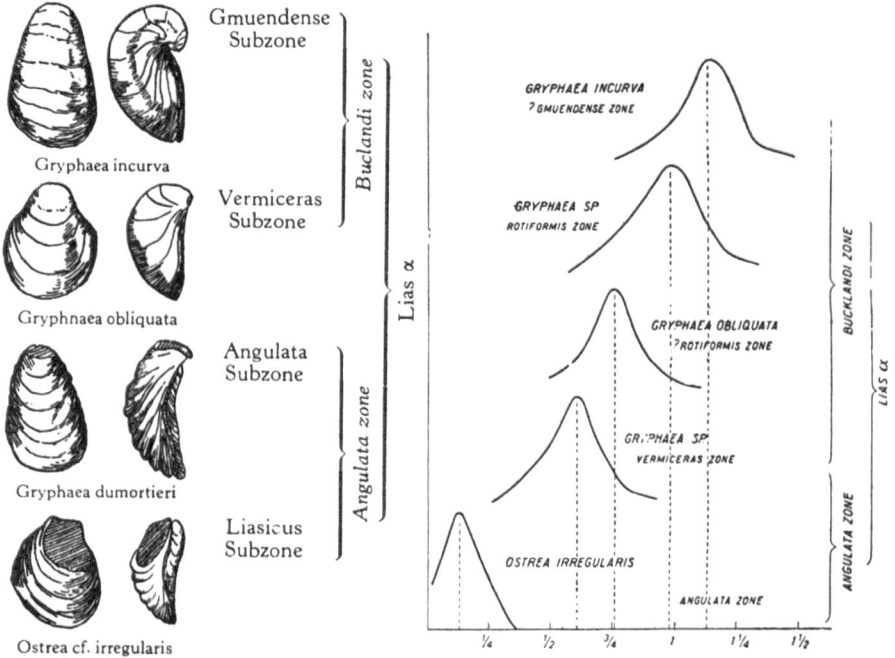

Fig. 95. Oysters belonging to the lineage Gryphaea incurva. (Adapted after A. E. Trueman).

Fig. 96. Graphs showing the progressive evolution of the degree of coiling in succesive populations belonging to the lineage Gryphaea incurva. (After A. E. Trueman).

$1^1/2$ to about 3 millimetres. The manner in which this variation is exhibited in time succession is shown by the three successive graphs of fig. 94 which are based on material collected from successively higher parts of the exposed strata (A, B, and C are based on material collected between 541–559 centimetres, 560–680 cm, and 681–854 cm respectively). The three crests of the upper curve (C) correspond with the three rib-types portrayed in the lower illustration above the word Anakosmoceras.

This example demonstrates the differentiation or "splitting" of a population into three separate populations, each of which corresponds to a new subspecies.

The next example concerns a collection of fossil oyster shells from Liassic deposits in South Wales and will show us how higher taxonomic categories (species and genus) evolve gradually from an ancestral population in the course of time.

As a matter of fact, two modifying tendencies were found by Trueman to be especially clear (fig. 95). One is a reduction of the flat area of attachment of the shell. Another is the progressive increase of coiling of the left valve from a flat oyster-like shell (Ostrea irregularis) to the incurved type, Gryphaea incurva. Let us focus our attention on the degree of coiling which in five successive horizons increased regularly from less than half a coil to one full coil and a half. If a large collection of shells is made from one particular level or horizon it is found that most individuals exhibit approximately the same degree of coiling, whereas the remainder are coiled to either a greater or a lesser extent than this. Their distribution expressed in a frequency curve shows a distinct crest the place of which is characteristic for a population of shells of that special horizon. At successive higher horizons, however, the degree of coiling exhibited by the greatest number of shells steadily increases. This phenomenon is expressed by a gradual displacement of the crest of the curve from left to right when proceeding from lower towards higher horizons of the Lower Lias (fig. 96).

This interesting example is important in more than one respect. For, in the first place it shows that at any one particular level the character of coiling exhibits a variation which can be expressed by a probability curve.

However, apart from the factor or factors which are responsible for this phenomenon another factor must have caused the progressive displacement of the crest of the curve in one special direction. This rectilinear and gradual change of one or more characters is called *orthogenesis*, a term which I shall use in a purely descriptive sense.

Moreover, the example shows a progressive change in the heriditary outfit of the successive populations. The connecting link between successive generations is a single cell, the fertilized *germ cell*. This cell divides and redivides repeatedly and so ultimately produces the body or *soma*. Only a few cells remain unmodified and give rise to new germ cells. If one of these cells becomes fertilized it may produce a new soma. The germ plasm thus forms a continuous chain of cells, from which new somas are produced and die. The germ cells are the material carriers of the hereditary characters, whereas the soma is the temporary physical expression of these factors. Biologists have shown that the hereditary factors are

extraordinarily stable. Evidently, however, they are not immutable —
as Weismann thought — if the element of geological time is taken into
consideration. Imagine a biologist who had the opportunity of experimenting with a great number of individuals belonging to the species Eotitanops gregoryi from the Lower Eocene of North America. Undoubtedly
he would have been impressed by the great stability of the heriditary
outfit of Eotitanops. However, if he had been able to continue his obser-
vations over a span of time of some 15 or 20 millions of years he would have seen how after countless gener-
ations the material offshoots of the germ cells changed from Eotitanops in the Lower Eocene to Brontops in the Lower Oligocene (fig. 97). He would have witnessed the phe-nomenon of orthoge-netic change of several characters.

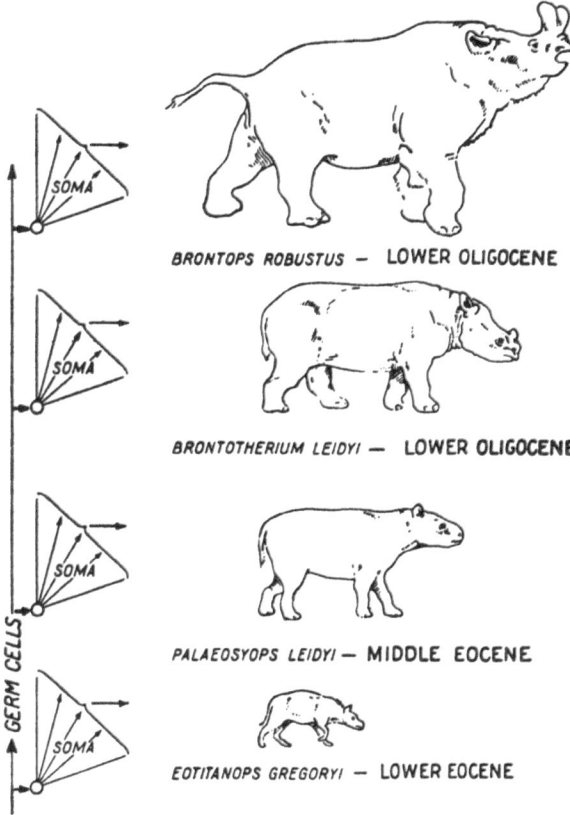

BRONTOPS ROBUSTUS — LOWER OLIGOCENE

BRONTOTHERIUM LEIDYI — LOWER OLIGOCENE

PALAEOSYOPS LEIDYI — MIDDLE EOCENE

EOTITANOPS GREGORYI — LOWER EOCENE

Fig. 97. Four stages in the evolution of the Titanotheria.
(Adapted after H. F. Osborn).

On closer inspection orthogenesis appears to be far from purely recti-linear evolution. The evolving lineage is made up of several lines that evolve only approxi-mately in one direction though not at constant rates and often with pronounced changes of direction and branching. Some of the evolving characters are positively
correlated others decidedly not. An organism is not a bundle of separately
evolving characters. Take one example, to which many others could
be added, viz. the evolution of the horse family (fig. 98). Eohippus, the
earliest-known ancestor of the horse was no larger than a small sized
and swift-footed dog. Its foreleg had four toes, the hindleg had only

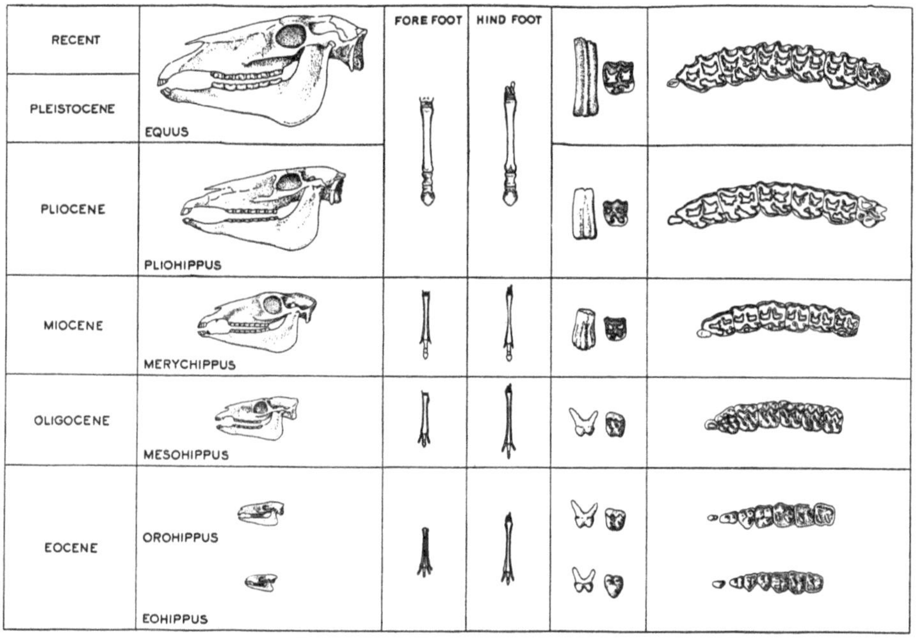

| | | FORE FOOT | HIND FOOT | | |
|---|---|---|---|---|---|
| RECENT | | | | | |
| PLEISTOCENE | EQUUS | | | | |
| PLIOCENE | PLIOHIPPUS | | | | |
| MIOCENE | MERYCHIPPUS | | | | |
| OLIGOCENE | MESOHIPPUS | | | | |
| EOCENE | OROHIPPUS | | | | |
| | EOHIPPUS | | | | |

Fig. 98. Main phases in the evolution of Equidae. (Adapted after H. F. Osborn).

three toes. The molars had short crowns bearing six cusps. The evolution of these primitive ancestors of the horse shows the gradual enlargement of all the parts of the body and its skeleton, a progressive specialization of the molars, molarization, which means change of the premolars into a dental battery, increase of the relative height of crowns (hypsodonty) and perfection of the coronal (lophiodont) pattern, a progressive reduction of the number of lateral toes and the adaptation of the only toe that is left to the practice of running. These evolving characters form a harmonious whole, though some of the evolving characters are intimately linked while others are not.

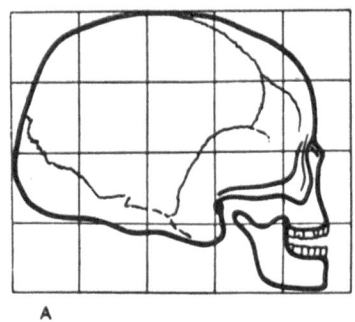

A

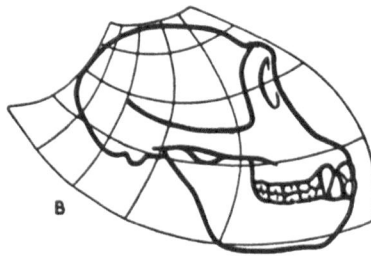

B

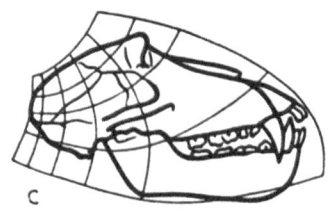

C

Fig. 99. A, outline of a human skull inscribed in Cartesian coordinates; B ,outline of skull of chimpanzee; C, outline of skull of baboon. (After D'Arcy W. Thompson).

Fig. 98 is a simplified representation showing only the main evolutionary trends of the horse family in a very schematic manner. In reality we are dealing with a very complicated story, consisting of a dozen branches evolving with many differences and irregularities of trend. Some characters probably evolved continuously, others not; some were highly correlated with each other, some were slightly correlated, others evolved independently.

A well-coordinated unity displayed by the evolution of a lineage strongly reminds one of the well coordinated unity displayed by a living creature during its individual development. More than that, it seems very probable that both phenomena are controlled by the same, still problematic factors. If the correlation between several evolving characters is not strongly positive the whole complex of characters becomes, as it were, plastically deformed. For the sake of demonstration one might inscribe them in a network of Cartesian coordinates which subsequently is gradually deformed. In this way d'Arcy Thompson inscribed the outline of a human skull in rectangular coordinates and by harmonious and congruent deformation of the coordinate network transformed it into the skull of a chimpanzee, and by a similar though intenser degree of deformation into the skull of a baboon (fig. 99).

Changes of direction and "tempo" in evolution are phenomena of frequent occurrence. An increase in evolutionary rate is termed *anagenesis*, the reverse — involving what may be called a regression — is termed *katagenesis*.

Changes in evolutionary rate have been established in evolving populations and even in greater assemblages of organisms consisting of at least a hundred different species. The average rate at which the reef corals evolved in the East Indies since Miocene times can be estimated from the percentage of still living species found in Pleistocene and Tertiary deposits. The evolution of the reef corals is graphically represented (fig. 100) by a curved line showing two sharp upturns, one in the Upper Miocene, another in the Pleistocene. Comparison with a similar curve constructed for the

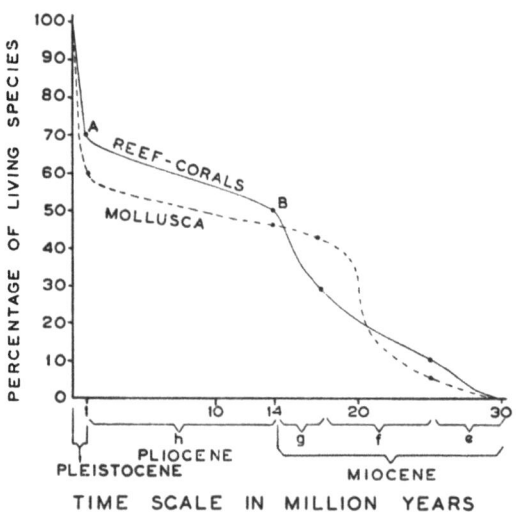

Fig. 100. Evolution of reef corals and molluscs in the East Indies since Miocene times. Percentage of living forms plotted against the absolute time-scale.

evolution of the Mollusca of the same area shows also two accelerations though not coinciding with those of the reef corals [1]).

A certain character appears at a given stage of the individual (ontogenetic) development of an organism. If this phenomenon is studied in subsequent geological time-levels of a lineage it can be found either to appear constantly at the same stage of individual development or to exhibit accelerations or retardations with geological time. Evolutionary acceleration of this sort is termed *tachygenesis*. Fig. 101 shows three spires of Procerithium shells from Lower Liassic deposits in England. The whorl number is indicated by II–IX according to their ontogenetic order of development; (a), (b), and (c) represent three subsequent stages of evolution. All three show one spiral costa (ridge-like marking) in whorl II, while there are two such spiral costae in whorl III. In more adult

[1]) Compare Simpson in "Genetics, Paleontology and Evolution", p. 215.

whorls the number of spiral costae increases. Moreover, additional ornamentation of the shell is exhibited by ridge-like markings running more or less parallel to the axis of the shell (axial costae). When both

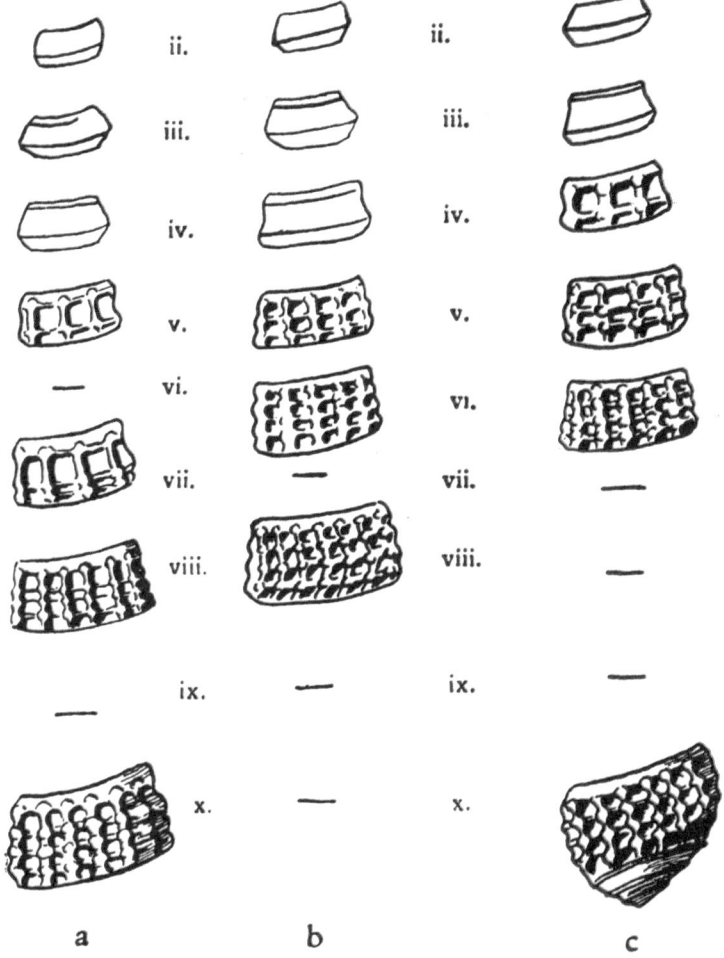

Fig. 101. Tachygenesis in the evolution of Procerithium from the Lower Liassic of England. (After McDonald and Trueman).

spiral and axial costae are present and equally developed a cancellated appearance of the shell is produced. Fig. 101 shows that the two-spiral ornamentation is ontogenetically exhibited longer in (a) than in (b) and (c). The axial costae appear earlier in (c) as opposed to (a) and (b). The same holds good regarding the cancellated stage with four spiral costae.

The reverse phenomenon exhibiting an evolutionary retardation in the appearance of ontogenetic stages, is termed *bradygenesis*. Fig. 102 portrays two shells belonging to the katagenetic part in the evolution of Procerithium. In (b) the ornamentation of two spiral costae is ontogene-tically longer lasting than in (a); axial costae in (b) appear half a whorl later than in (a), and the cancelled pattern is not so well developed in (b) as compared with (a).

Tachygenesis and bradygenesis are intimately interwoven with other remarkable evolutionary phenomena. To elucidate this let us first mention that the origin of new characters and the persistence of ancestral ones can occur in two ways, one of which is the exact reverse of the other.

Specimens carefully collected from successive zones have shown that Cyclolituites in which the whole shell was coiled arose from Rhynchor-toceras, which was straight or slightly curved. Ancistroceras and Lituites are intermediate between these two extremes in shape as well as in the time of their occurrence (fig. 103). The condition which characterized Cyclolituites was first manisfested in the youth of Ancistroceras and Lituites. This principle of anticipation is termed *proterogenesis*.

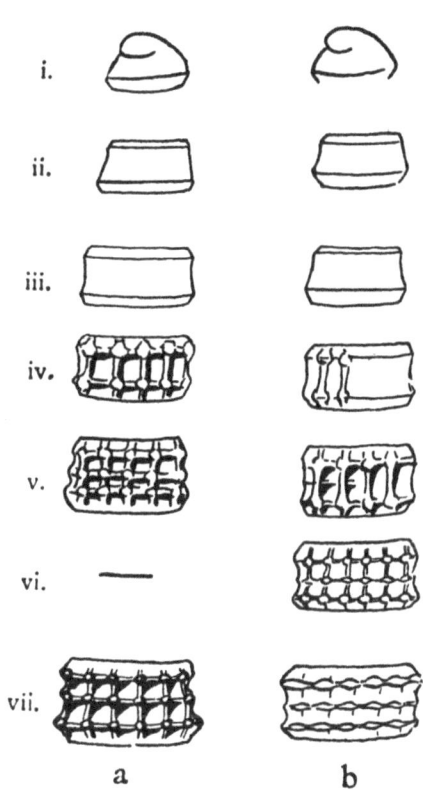

Fig. 102. Bradygenesis in the evolution of Procerithium from the Lower Liassic of England. (After McDonald and Trueman).

Quite the reverse is shown in the evolution of other animals. There seems not the slightest doubt that Cycloclypeus evolved from Heteroste-gina and that this Foraminifer evolved from Operculina. In this case the new characters are exhibited by the adult stage of the individual. This phenomenon, termed *deuterogenesis*, is a complete reversal of the sequence of events occurring in proterogenesis. On the other hand the ancestral characters are recapitulated in a young stage of individual development. Heterostegina starts growing as an Operculina, Cyclo-clypeus begins its development as an Operculina and then passes through the stage of Heterostegina (fig. 104). The young stages recapitulate, as it were, the characters of their ancestors. This phenomenon is termed *palingenesis*.

Moreover, the evolution of Cycloclypeus since the upper part of the Tertiary shows tachygenesis which, in combination with palingenesis, results in the skipping of certain ancestral stages termed *lipopalingenesis*.

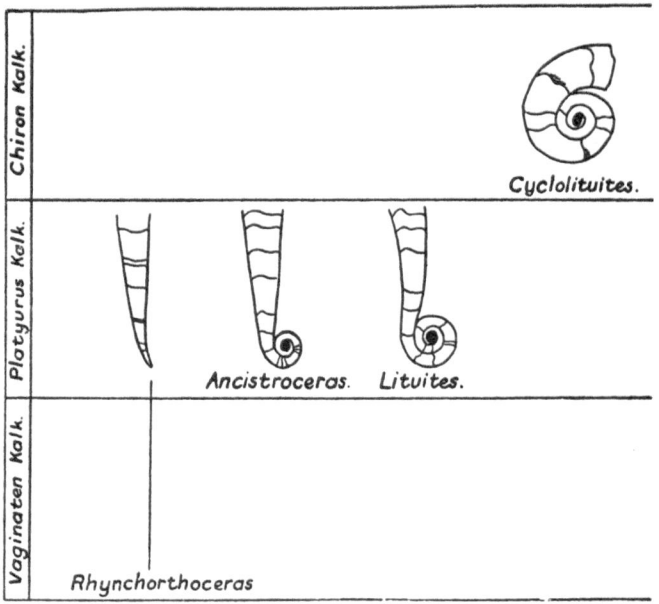

Fig. 103. The phenomenon of proterogenesis in the evolution of Lituitidae
(After Schindewolf, from H. H. Swinnerton).

To make this principle clear it must be mentioned firstly that fig. 104 represents an early species of Cycloclypeus in which both Operculina and Heterostegina stages may be recognized (palingenesis). In species belonging to successively higher horizons the circularly arranged chamberlets occupy an ever-increasing portion of the shell. They shift back to successively earlier growth stages (tachygenesis), until at last they occupy nearly the whole shell with the exception of a small central area (fig. 105). The Heterostegina stage undergoes progressive reduction, while the Operculina stage is eliminated entirely (lipopalingenesis).

To complete our short review of the different aspects of the complicated phenomenon which is designated by the term orthogenesis, we shall return now to the evolution of the Titanotheria.

The evolution of this extinct group of Titanotheres has been studied by Osborn in great detail. As in most cases, it appears that we are not dealing with one lineage but with several separate lineages which evolved independently, though showing similar trends in their progressive charac-

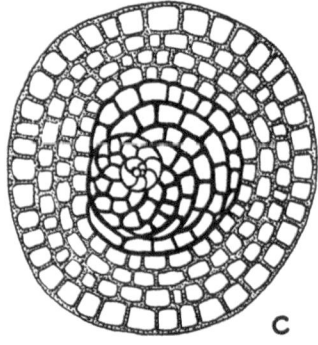

ters. Thus the horns in the Titanotheres arise and evolve independently at different times in different lineages which all descended from the same primitive ancestral population. The time of their origin is indicated by the black dots in the upper part of fig. 106, the place where the horn starts its evolution is indicated by the letter H in the lower part of the illustration. In the same way the larger number of cusps on the teeth as well as the change of proportions both in the skulls and metapodials have arisen quite inde-

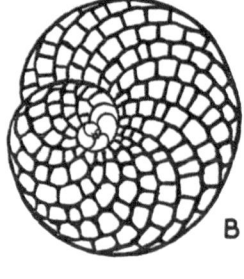

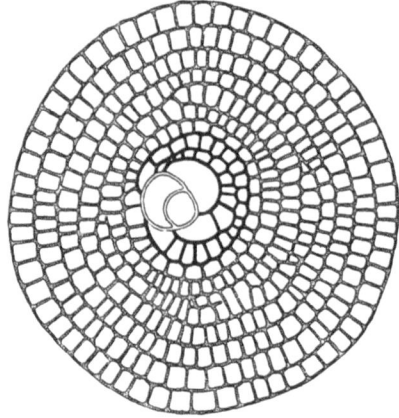

Fig. 104. The phenomenon of palingenesis of adult stages demonstrated by the sequence Operculina (A), Heterostegina (B), and Cycloclypeus (C), three Foraminifera from Upper Tertiary strata of Java.

Fig. 105. Cycloclypeus guembelianus from the Pliocene of Java, showing palingenesis, tachygenesis and lipopalingenesis.

pendently of each other in different phyla. This remarkable phenomenon is termed *homomorphy*. Another example of homomorphy is portrayed in fig. 17 (p. 26) which not only shows the orthogenetic evolution of numerous lineages in the family Mastodontoidea but also the great amount of parallelism (homomorphy) of similarly specialized organs like the shortened jaw, the increasing complication in the structure of the molars, the evolution of a trunk and the increase of body size, independently produced in different lineages.

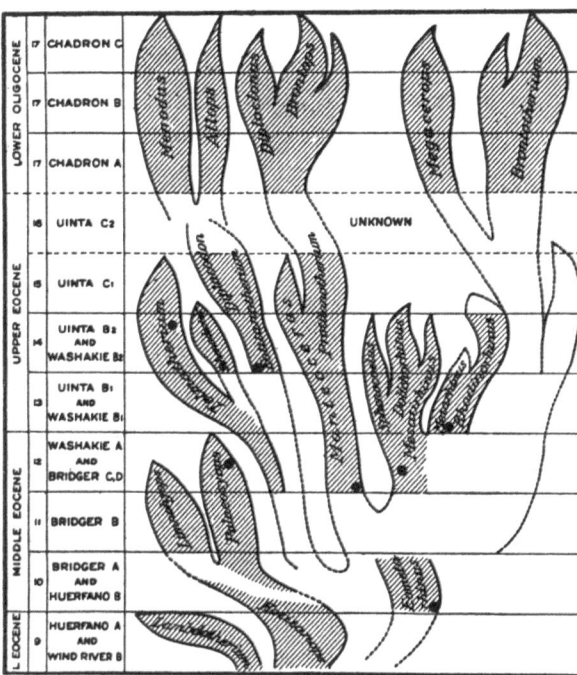

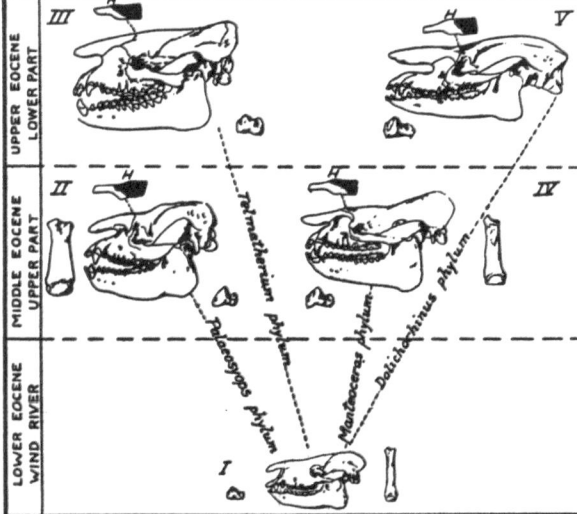

In these examples the whole ancestral stock evolves into new species and genera. Sometimes, however, part of the ancestral types remains unchanged whereas another part gradually changes into one or more different types. Fig. 107 shows the evolution of three lineages of Bryozoa belonging to the genus Stomatopora. In (a) there has been hardly any change, whereas in (b) and (c) the angle between branches decreases and the bases of adjoining branches tend to fuse.

The evolution of lineages (b) and (c) is one problem, lack of evolution of lineage (a) is another problem. It is often seen that the more primitive ancestral group remains unaltered or changes little during long periods, whereas the offshoots show much stronger evolution and specialization.

The simultaneous occurence of organisms which are primitive and

Fig. 106. Parallel evolution (homomorphy) in the family of Titanotheria. (After H. F. Osborn).

which evolved at a low rate as compared with specialized organisms of an allied group, is a wide-spread phenomenon. It is illustrated by the occurrence in our present flora and fauna of organisms which, in a broad sense, represent successive steps of evolution. Primitive algae and highly specialized plants are found at the same time, Protozoa are living besides Metazoa etc. etc.

These are examples of the persistence or low-rate evolution of primitive types which did not evolve far from the ancestral stock from which other specialized and high-rate lines differentiated.

It appears as if slowly evolving lines are almost immortal. On the other hand lines which evolve at high rate into highly specialized types are as it were rushing toward extinction. Yet there are very remarkable exceptions to this general rule. This brings us to another type of low-rate evolution or "arrested evolution" as Ruedemann termed it. It is

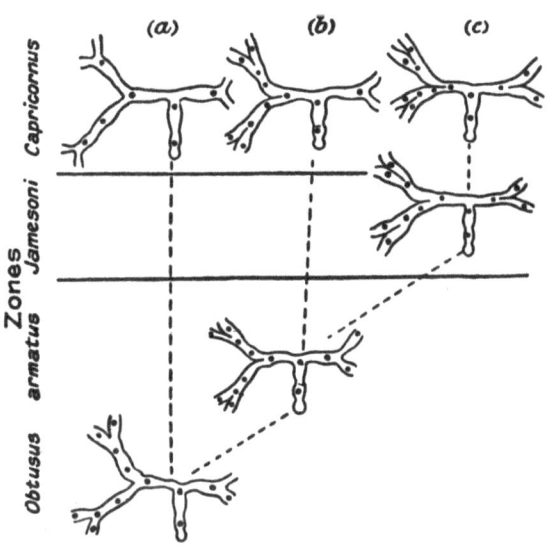

Fig. 107. Evolution of several lineages of Stomatopora. (After Lang, from H. H. Swinnerton).

represented by organisms which reached a strikingly complicated stage of specialization and then persisted unaltered or with hardly any change.

Among invertebrates Lingula is a well-known example of an animal which has remained unchanged during the 500 million years which elapsed since Cambrian times (fig. 108), which means that during some 50 to 100 millions of generations the hereditary characters of the germ cells remained stable. Other well known examples are Nodosaria, Saccamina, Limulus (fig. 108), Nautilus, Cidaris, Pleurotomaria, Crania, Rhynchonella, Terebratula, Ceratodus, Chelone, Sphenodon, Sequoia, Araucaria and Ginkgo. Once more I should like to underline the fact that apart from the remarkable phenomenon of evolution several groups of animals and plants do not exhibit signs of evolution during long geological periods.

High-rate and low-rate or arrested evolution are intimately interwoven phenomena in the history of most of the higher and more complex taxonomic categories. The evolution of Brachiopods and Cephalopods furnish striking examples in this respect. It is a fascinating problem why all this happened exactly as it actually did happen.

One of the most remarkable and problematic phenomena in organisms is their becoming adapted to their organic and inorganic environment.

It is often found, as in Osborn's research on the Proboscidea, that a newly evolving group diversified and produced types which became adapted to very different habitats "from the purely aquatic and amphibious habitat

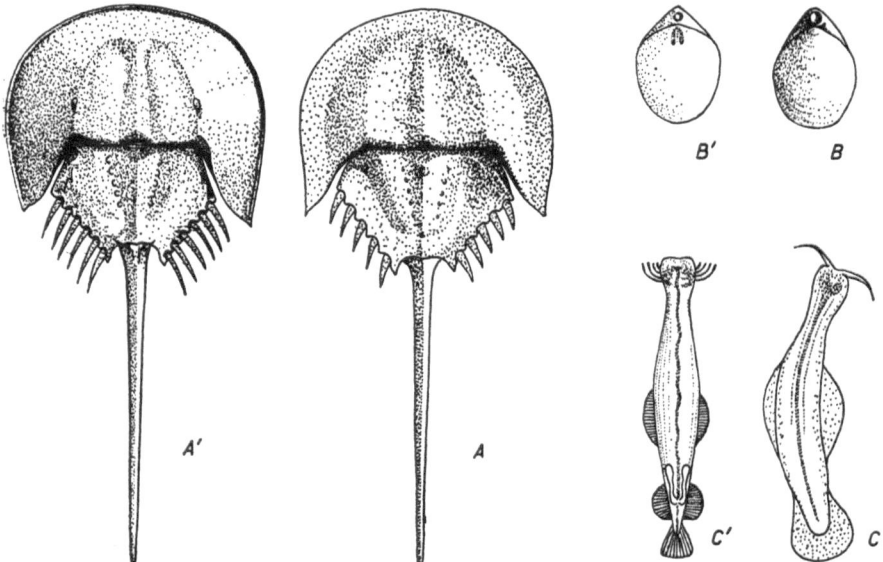

Fig. 108. Examples of persistent types during long geological periods, A, Limulus from the Upper Jurassic of Solnhofen and A' recent specimen of Limulus; B, Terebratula subsella from the Upper Jurassic of the Harz Mountains, B', Terebratula vitrea living in the Mediterranean; C, Amiskwia from the Mid-Cambrian of Colombia and C' Sagitta a recent marine worm.

of the Moeritheres, through swamp living, river border living, shallow lake border living, savannah living, forest living, tundra living and desert living, from equatorial to boreal latitudes and from sea-level to elevated mountain habitats; for example, the Andean mastodont".

This sort of adaptation was termed *adaptive radiation* by Osborn. Moreover, the organism shows an adaptive response of each particular part of the body to the mechanical and structural problems set by its environment. It is a well-known fact that the final products may consist of remarkably well adapted organisms, and that convergent types may arise in such widely different groups of organisms as fishes, reptiles and mammals (fig. 109).

Evolution, therefore, clearly shows the continual influence of the environment on the endless chains of living beings resulting in changes of proportion and function as well as in the originating of new and distinct adaptations, which have become part of the hereditary equipment of the organism.

It goes without saying that if an organism became especially well adapted to a certain environment it must have been pre-adapted to a certain extent, making it viable in that environment.

Some examples of specialization have led to the conviction that once a special direction of evolution has started it tends to continue in that direction farther than the bounds of utility. The result may be disadvantageous or even lethal.

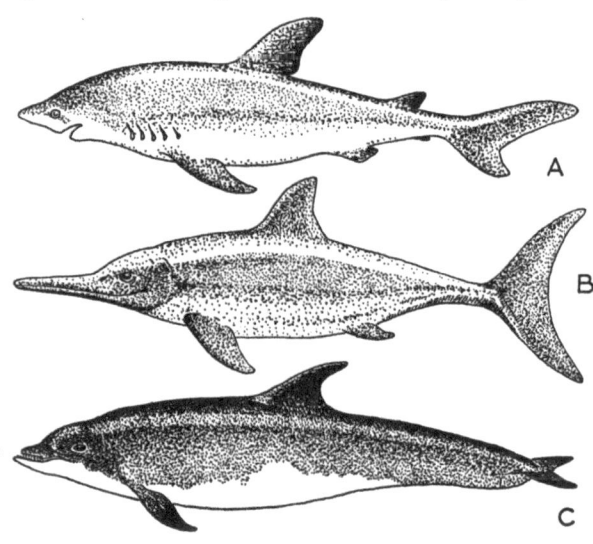

Fig. 109. Convergent adaptations as shown by a shark (A) an Ichthyosaurus (B), and a dolphin (C).

A few examples are shown in fig. 110. The strongly curved tusks of Babirussa cannot possibly be considered a useful weapon of attack or defence. Possibly the enormous growth of the antlers of the extinct Irish stag, Megaceros hibernicus, even became fatal to that species, because of their enormous weight and inconvenient dimensions. In general there is a reduction of adaptability with increase of specialization. Moreover, specialization, i.e. extreme adaptation to a certain habitat may involve reduction of certain other characters and organs which are not employed and are therefore no longer of any use. A well-known example is the reduction of the pelvic bones of aquatic mammals like the Sirenidae. Another example is the reduction of the wings in some birds like the ostrich and the Dodo (fig. 111).

It is not thinkable that organs which have been reduced to rudimentary remains or which have been lost entirely will ever become regained in the course of further evolution. In case of return to a mode of life in which a lost organ would be useful, other parts of the organism are — if

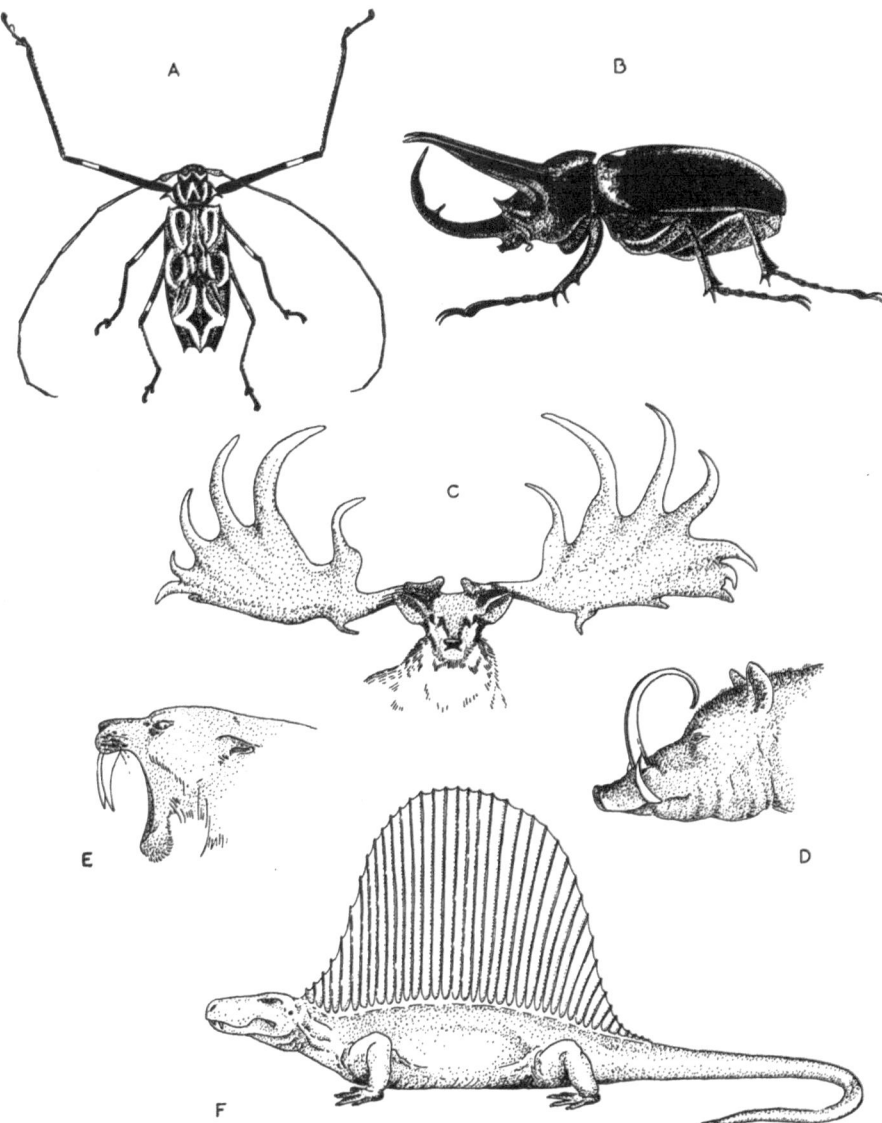

Fig. 110. Examples of organisms in which evolution of certain characters surpassed the bounds of utility (hypertrophy). A, the forelegs in Acrocinus longimanus from Brasil; B, the horns in Chalcosoma atlas from Java; C, the antlers in the extinct giant Irish deer Megaceros hibernicus from the Pleistocene of northwestern Europe; D, the curving of the tusks in Babirussa from Celebes; E, the length of the canines in the saber-tooth Smilodon from the Pleistocene of California; F, the elongation of the dorsal spines in Dimetrodon from the Permian of Texas.

possible — evolved or deformed so as to take over the function of the lost organ. This is the so-called law of irreversibility or rule of Dollo, who gave several examples substantiating this opinion. In many cases, however, the organism has become specialized or over-specialized to such a degree, that hardly a possibility of evolution in the same or any other direction seems left. These creatures are doomed when conditions of life change. According to Decugis our present fauna bears many unmistakable signs

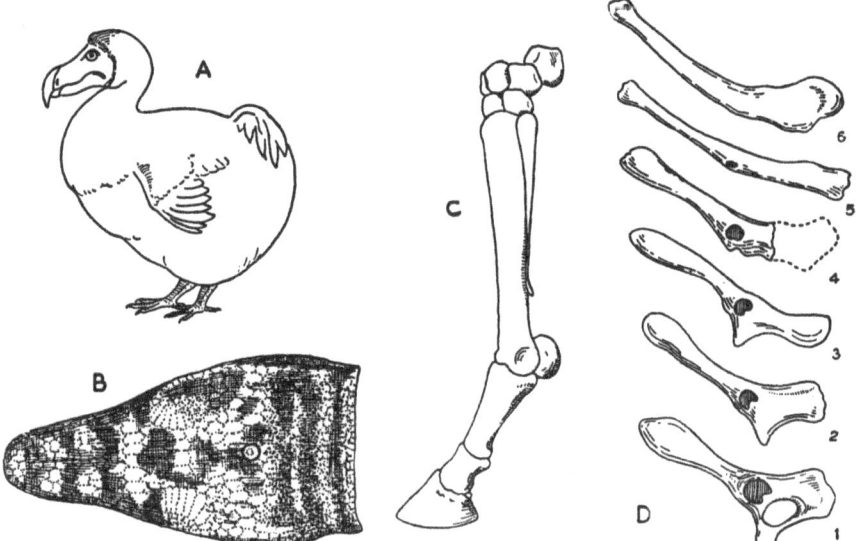

Fig. 111. Rudimentary organs: A the wings of Dodo (Didus ineptus); B the pineal eye of Varanus; C reduction of lateral digits in the horse; D reduction of the pelvic bone in the family Sirenidae, 1 Eotherium (Mid-Eocene), 2 Eosiren (Upper Eocene), 3 Halitherium (Oligocene), 4 Metaxitherium (Miocene), 5 Halicore tabernaculum (Recent), 6 Halicore dugong (Recent).

of senility. Broom is even more pessimistic. In his opinion evolution has attained a dead level, whereas Leconte du Nouy holds that mankind represents the only species which is capable of still greater changes, and further evolution. We shall return to this question later on (p. 195).

Periods of great changes in the earth's physical and climatological aspect must have been fatal to many organisms which were strongly adapted to special conditions of life. Yet in this respect the influence of the environment should not be overrated. It is, for example, difficult to see why creatures like Ichthyosaurus and Mosasaurus could not find a suitable habitat in one of the seven seas towards the end of the Mesozoic. Was their extinction due to "internal" factors? Should we distinguish something like birth, youth, adultness, old-age and death in the evolution of a lineage?

Answering these questions in the affirmative would mean entering the domain of pure speculation. Let us therefore draw our attention to another aspect of evolution.

The graphs of fig. 94 illustrated how a population is fragmented into three isolated lines of descent. It differentiates into what a taxonomist would label as subspecies.

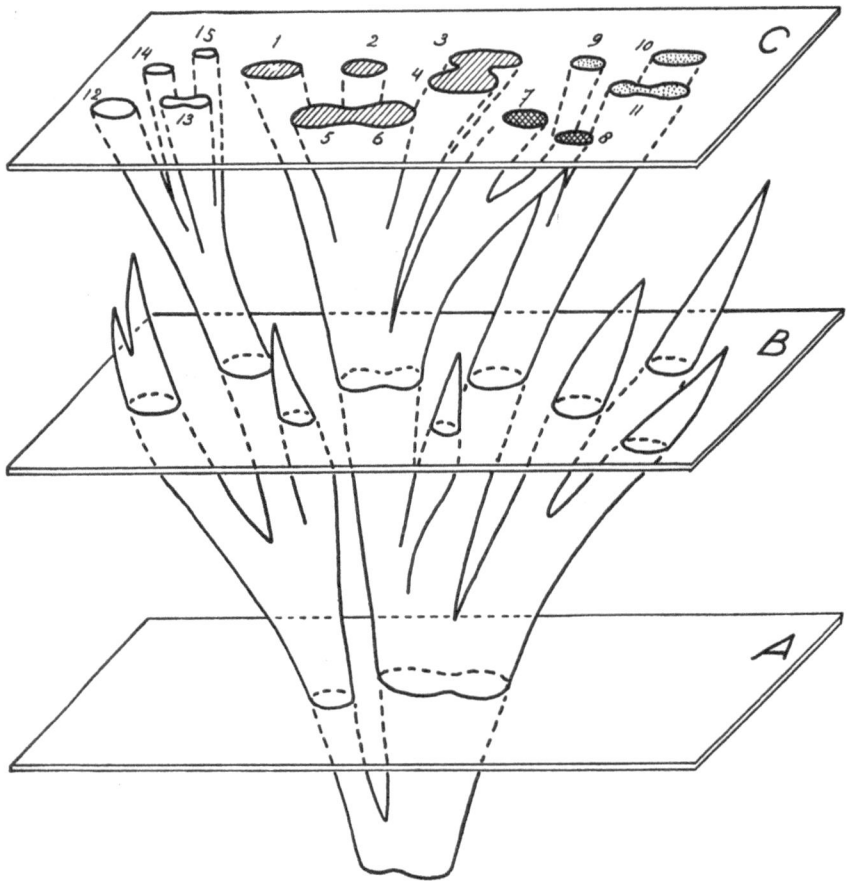

Fig. 112. Schematic representation of the evolution of 15 subspecies, species and genera.

The example of Gryphaea incurva (fig. 95 and 96) demonstrated the gradual change from one species to another species and so on, the final product being definitely classified as a different genus.

Gradual and continuous change and differentiation of populations is considered by some authors as the normal and basic process in evolution.

They hold that the longer the span of time involved the greater the differ-
rence from the taxonomist's point of view. In a diagrammatic manner
this opinion is expressed in fig. 112, A, B, and C representing three
successive, though arbitrary time-levels. In time-level C six species, 1–6,
are considered as belonging to one and the same genus, 1 and 2 being two
distinct and well defined species, 3 and 4 as well as 5 and 6 intergrade.
The species 7 and 8 are thought to belong to a different genus, 9, 10 and
11 represent again different genera. Finally 12, 13, 14 and 15 are so different
that a taxonomist would classify them in a higher systematic category for
example in a different family.

Fig. 112 expresses the viewpoint that the discontinuities between
species and genera as seen by the taxonomist in one horizontal time-level
can arise by a continuous process of gradual change and differentiation
vertically, that is in geological time.

Remarkably enough, however, paleontological data reveals numerous
examples of the gradual and continuous differentiation into subspecies,
species and genera, but they seldom give an adequate answer to the
important question how the greater changes occurred that gave rise to
quite new types and to some of the higher categories (phyla, orders, classes,
families). For several decades this was thought to be inherent in the paucity
of the fossil record itself. To be sure, several important finds have been
made revealing the former existence of creatures which for a long time were
considered as missing links. Examples are Archaeopteryx and Archaeornis,
Ictidosaurus, Ichthyostegale and Ichthyostegopsis, Eunotosaurus, Protoba-
trachus, Hornea, Rhynia and Asteroxylon, Pithecanthropus and many
others.

Still it remains a striking fact that something like a continuous orthogenetic series is conspicuously absent at most of the critical points of evolution. It has therefore been suggested by some authors that instead of
ascribing this negative result to the deficiency of the paleontological
record we should see it as a salient feature in the evolution of the world
of organisms, as a separate problem of evolution which calls for explanation. Several phyla, orders and families appear suddenly in the record,
displaying a sharp discontinuity in the pattern of their characters and
organization as compared with earlier groups of organisms.

It was thought plausible to ascribe a certain role to the occurrence of
sudden discontinuous changes of characters or groups of characters.

Discontinuity is a phenomenon which biologists have observed in the
recent fauna and flora and which is known under the term saltation
(= mutation of Hugo de Vries). More than that, saltations have been

provoked artificially, e.g. by X-ray treatment of the little fly Drosophyla.

It is difficult to ascertain similar discontinuous changes of characters in paleontological material. Many discontinuities in the fossil record may only be apparent due to the very common phenomenon of discontinuity in sedimentation. An example is shown in fig. 113. The number of inner

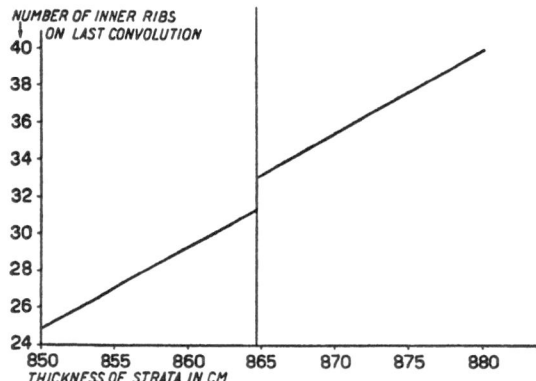

ribs on the last convolution of the Jurassic ammonite Zugokosmoceras increased steadily from about 25 to about 40 during a span of time expressed by the deposition of 30 centimetres of strata, from 850 to 880 centimetres in the section of Oxford clay studied by Brinkmann.

Fig. 113. An apparent saltation in a lineage of Zugokosmoceras due to a break in the normal sedimentation.

The graph shows, however, an abrupt offset at 864.5 cm. The place of the apparent discontinuity corresponds with a stratigraphically recognizable break in the normal deposition of sedimentary material.

Evidently, a certain span of time is not represented by sediments, involving a hiatus in the fossil record and therefore an apparent but not a true discontinuity in the evolution of Zugokosmoceras.

In other cases, however, the general incompleteness of the fossil record as well as the evident time gap between successive paleontological data seems insufficient to account for the evolutionary advance without postulating either an abnormally high rate of evolution or saltations.

Saltations of the type known to biologists can have occurred in any part of the complex evolutionary process. Let us recall the evolution of Cycloclypeus (fig. 104 and 105). The number of circularly arranged rings increased progressively with geological time. Now, very probably this increase in number was a discontinuous, saltatory process.

Possibly, however, saltatory changes occurred on a large scale during certain stages of evolution. It is very difficult to understand how gradual evolution could have resulted in such drastic changes as are needed to explain the case of an animal that passed for instance from a land habitat into an aquatic habitat. Pronounced, radical and pre-adaptive changes in physiological and mechanical systems were involved in these evolutionary phases. It has therefore been suggested by Schindewolf that the main new types from which a new lineage or phylum started to evolve originated suddenly, saltatory and according to the principle of protero-genesis. Without gradual changes the new organism is supposed to have jumped on the stage in full harness, more or less like Pallas Athene jumped out of the head of Zeus. Goldschmidt introduced the term macro-mutations. To avoid misunderstanding *macro-saltation* would perhaps be a better term.

Thus, we might distinguish two fundamentally different modes of evolution. One is represented by the gradual changes shown by the origin of subspecies, species and genera, by branching of populations, orthogenesis and homomorphy. It is known from numerous examples revealed by paleontological data concerning Vertebrates and Invertebrates, We might include in it saltations of the normal type. The other is the macro-saltation which becomes the starting point of series of more or less gradual changes. The supposed process of a subsequent macro-saltation is thought to give rise to proto-types of new taxonomic units of relatively high rank. The new types must have been pre-adapted — in their characters, functions and instincts — to certain minimum conditions of life set by the organic and inorganic environment.

Moreover, epochs of macro-saltations are thought to have alternated with periods of adaptive radiation and a more gradual orthogenetic evolution. Epochs of macro-saltations were held to coincide with epochs of great revolutions in the geographic and climatic conditions of our globe. Possibly periods of great changes in the physical aspect of the earth acted directly or indirectly through the climate. Others suggested intensity of solar radiation or the increasing action of cosmic rays which stimulated somewhere the pulse of life and gave rise to a sort of explosive origin of new types at random. Possibly many of these creatures were simply deleterious monstrosities unable to procreate or to evolve. But perhaps some were "hopeful monsters" which formed the starting point of one or more further evolving lineages.

The hypothesis of macro-saltations seems to me a too idealistic product of pure speculation. On closer inspection the proclaimed macro-saltations appear to be varying as to their sequence, duration, time of occurrence and magnitude of effected evolutionary change. Moreover, so many transitional forms are known between reptiles and mammals that drawing a dividing line between these two phyla would be an arbitrary procedure. And the same applies to the transition between fishes and primitive tetrapods.

Moreover, finds of "missing links" warn us not to jump too readily to conclusions based on the "negative evidence" of the paleontological record. The presence of fossil remains is always due to the accidental concurrence of exceptional circumstances. Let us recall the famous find of two primeval fossil birds in the Upper Jurassic limestone of Solnhofen in the years 1863 and 1877 respectively. Not a single specimen has been discovered during the years which have elapsed since 1877 though the quarries have been intensively exploited and explored. The two specimens, the only remains of Jurassic birds, were classified as belonging to two different genera, termed Archaeopteryx and Archaeornis respectively. Of course, it would be absurd to suppose these were the only two sorts of birds living in Upper Jurassic times as well as the only birds for many tens of millions of years in earlier and later times. Without the good luck of finding these two extremely interesting "missing links" some authors would undoubtedly have formulated the theory of a spectacular macro-saltation giving origin to the bird phylum in Cretaceous times. Yet now Schindewolf would have us believe that at one time in the Jurassic an ancestral type of bird sprang chirping from a reptile's egg.

However, even if the macro-saltation concept is abandoned entirely, considerable evidence is in favour of the probability of the occurrence of explosive phases of high rate or discontinuous changes, for instance an accumulation of normal mutations or saltations in certain stages of the evolutionary process. As Simpson said, it appears "that major transitions do take place at relatively great rates over short periods of time and in special circumstances". Simpson designated this process as *quantum evolution*. In his own words: "The term is applicable in situations in which sub-threshold actions produce no reactions but super-threshold actions produce reactions of definite (not necessarily equal) magnitude (this magnitude being strictly the quantum involved). For the sake of brevity, the term "quantum evolution" is here applied to the relatively rapid shift of a biotic population in disequilibrium to an equilibrium distinctly unlike an ancestral condition. Such a sequence can occur on a relatively small scale in any sort of population and in any part of the complex evolutionary process".

Fig. 114 summarizes the different patterns of evolution in a diagrammatic manner; see also Table III, pp. 174–175.

Perhaps Lull was right when he wrote: "changing environmental conditions stimulate the sluggish evolutionary stream to quickened movement". Concerning this point, however, we are in need of much more carefully gathered evidence. For the moment let us recall one of the examples mentioned in chapter I (p. 29). The only two periods of strong differentiation of the flora into several botanical provinces coincided with epochs of the greatest revolutionary changes in the physiographic aspect of the earth's surface since the appearance of plant life (fig. 18). They were two periods of tremendous mountain-building and exceptionally

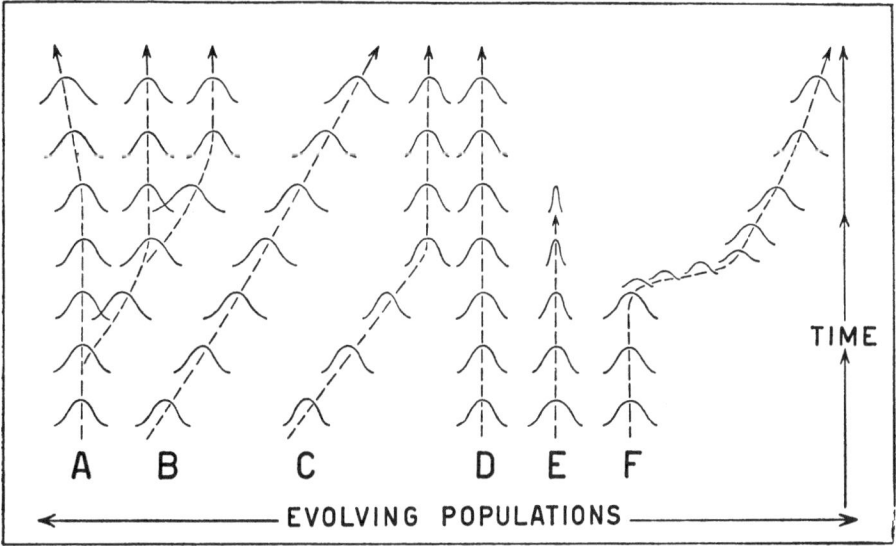

Fig. 114. Schematic and tentative representation of six patterns of evolution. A, origin of subspecies, compare fig. 94; B, orthogenesis; C, arrested evolution; D, persistent type; E, specialization and extinction; F, "explosive" evolutionary change, "quantum evolution". (Partly based on G. G. Simpson).

widespread glaciation. Conversely, in the intervening periods the land flora developed into what might almost be called a world-wide uniformity.

Phases of more gradual and continuous evolution are characterized by adaptation, adaptive radiation, specialization, in short by phenomena which reveal a marked interaction between organism and environment.

There was again a strong influence from the environment acting on the hereditary outfit of the organisms during the more rapid, discontinuous and explosive phases of evolution. It seems to me that this influence was twofold. One was the action which caused the latent accumulation of change in the hereditary outfit which subsequently appeared as pre-adapted saltations. The other was a sort of trigger effect. It suddenly released the pre-adapted changes in the hereditary outfit, which previously had accumulated in a latent state and which were apt to become phenotypic and genotypic evidence after latent accumulation had reached a certain quantum.

Another problematic manifestation of the endless succession of living systems is seen in orthogenetic lineages. Are orthogenesis and related phenomena controlled by an internal trend? If so what does that mean? Does there exist a definite limitation of the number of possible directions of structural modification inherent in the mechanisms of mutation and saltation? If random mutation and saltation is limited to a certain extent by such a process, to what extent then are the remaining possibilities still further limited and directed by natural selection?

These important questions embrace a whole series of fundamental problems which, as Simpson wrote "are so broad that they cannot hopefully be attacked from the point of view of a single scientific discipline". Evidently, the intimate cooperation is needed of biologists and geneticists, paleontologists and geologists, physicists and chemists. For the problems of evolution overlap all these different fields of research.

Serious attempts at uniting the results of genetics (including selection) and paleontology into a modern synthesis may be found in Simpson's *Tempo and mode in Evolution* as well as in the *Symposium* on evolution published by Jepsen, Mayer, Simpson *et al.* in 1949.

In the foregoing review we have seen some of the major features and problems of evolution. We learned that the evolving world of organisms is intimately interwoven with physico-chemical or "material" processes acting both in the organism and through its environment. The paleontological record greatly broadened our insight into the great diversity of phenomena in which living systems manifested themselves during the past 500 million years of earth history.

However, the crucial problem is: what are the processes characteristic of a living system as opposed to a non-living system, and what is their relation and interaction with the material processes in the living organism?

Finally, I want to emphasize the following two points in these considerations. Firstly, the picture of evolution seen by a paleontologist is essentially orderly. It seems therefore hardly possible to reconcile this picture with any theory on evolution based on a principle of pure chance, be it random mutation or random saltation. One is driven to the idea of directional mutation as well as to the concept of pre-adapted — not random mutations or saltations. Secondly, if phenomena of life and its evolution appear inexplicable on a basis of physico-chemical processes known from non-living systems, this should not *ipso facto* lead us to jump at metaphysical speculation. All that we can say for the moment is that living systems are controlled by processes which are essentially different from those controlling non-living systems. What these processes are and in what manner they are linked with and related to material processes in a living organism is one of the most baffling problems of science.

The foregoing very condensed review of some of the more salient features and problems of evolution has been kept free from speculation as much as possible. Let us now turn our attention to the major groups of theoretical considerations on the problems of life and its evolution.

### THEORETICAL CONSIDERATIONS

*Vitalistic theories.*

In the following pages we shall deal with some well-known theories concerning phenomena and processes which are characteristic of life. I hope nobody will blame me for looking again rather often through my geological spectacles.

I shall pass by the more remote past, e.g. the mediaeval theory of a life-force (*vis vitalis*), which, endowed with the qualities of the human mind, was designed to explain the processes and adaptations of the organism. This can be done the more readily because this opinion has been revived in the doctrine of the so-called neo-vitalism, which will be taken as the starting-point. For, revived as a reaction against "materialism" (which will be dealt with later), the neo-vitalistic hypotheses suppose that, both in the individual development and growth of a living creature and in the evolution of the world of organisms, there is a directing and regulating creative principle at work. In this respect it makes no difference whether the old *vis vitalis* is resurrected, whether new theories are propounded concerning the memory and mental power of living cells, or whether, in any other way, there is added to the physico-chemical processes in the living organism the operation of some other entity which is considered to be characteristic of a living system. This directing vital activity was idealized and personified in the likeness of a product of man's creative urge. This factor is common to many vitalistic hypotheses: it is thought that in living creatures one can detect a force which is directed to an end, — an event following a pre-arranged or well thought out plan. This supposed directing force is designated by the words *teleology* or *finality*. The wing of an albatross, the curved beak of an eagle, the webs of a duck, were so developed in order that the albatross could weather the storm, the eagle could rend the goat, the duck could become a swimming bird, and so on. They are organs which have grown in the structural pattern of the body together with the corresponding instincts according to a pre-meditated plan, — they are designs or inventions. According to Bergson this is also the inevitable background of Lamarckism which considers adaptation mainly as a reaction of the

TABLE III. *Synopsis of evolutionary phenomena.*

| Paleontological aspects of evolution |
| --- |

I. "Gradual and continuous" evolution in lower taxonomic units (subspecies, species, genus).
   A. Differentiation of population in 3 subspecies (fig. 114, A).
   B. Orthogenetic transformation into new species and genus (fig. 114, B).

   C. Some phenomena in "rectilinear" evolution.
     1. Change of direction; anagenesis and katagenesis.
     2. Change of evolutionary rate ("tempo").
     3. Ontogenetic acceleration, tachygenesis
     4. Ontogenetic retardation, bradygenesis
     5. Origin of new characters: a, proterogenesis
                             b, deuterogenesis
     6. Recapitulation of ancestral characters, palingenesis
     7. Skipping of ancestral characters, lipopalingenesis
     8. Parallel evolution, homomorphy

     9. Persistence of ancestral type (low rate evolution), (fig. 114, D)
    10. Arrested evolution (fig. 114, C)
    11. Monophyletic origin
    12. Polyphyletic origin
    13. Adaptation, specialization, adaptive radiation
       a. Convergent types (influence of the environment)
       b. Evolutionary "inertia" $\alpha$ hypertely
                      $\beta$ dystely
                      $\gamma$ ately (rudimentary organs)
       c. Irreversibility
       d. Senility, extinction (fig. 114, E)

    14. Saltation
    15. Pseudo-saltation
II. "Gradual and continuous" evolution in higher taxonomic units (familia, classis, ordo, phylum)
III. Discontinuous, "explosive" evolution
   A. Theory of macro-saltations ("hopeful monsters")
   B. Theory of quantum-evolution (pre-adapted saltations or mutations; influence of environment: latent changes in germ plasm, trigger effect). — (fig. 114, F)

| Examples | Illustrations |
| --- | --- |
| Anakosmoceras gulielmii | Fig. 94 |
| Ostrea → Gryphaea | Fig. 95 and 96 |
| Eotitanopos → Brontops | Fig. 97 |
| Eohippus → Equus | Fig. 98 |
| | |
| Reef corals and Molluscs in Indo-Pacific | Fig. 100 |
| Procerithium | Fig. 101 |
| Procerithium | Fig. 102 |
| Lituitidae | Fig. 103 |
| Heterostegina, Cycloclypeus | Fig. 104 |
| Heterostegina, Cycloclypeus | Fig. 104 |
| Cycloclypeus guembelianus | Fig. 105 |
| Titanotheria | Fig. 106 |
| Mastodontoidea | Fig. 17 |
| Stomatopora | Fig. 107 |
| Lingula, Terebratula, Limulus | Fig. 108 |
| Ammonitoidea | |
| Ceratitoidea | |
| | |
| Ichthyosaurus, dolphin, shark | Fig. 109 |
| Babirussa, Dimetrodon | Fig. 110 |
| Megaceros hybernicus | Fig. 110 |
| Sirenidae, Dodo | Fig. 111 |
| Dinosauria (pelvic bones, Dollo) | |
| Ichthyosaurus, Mosasaurus, | |
| Dinosaurs | Fig. 116 |
| Cycloclypeus | Fig. 104 and 105 |
| Zugokosmoceras | Fig. 113 |
| Reptilia → Mammalia | |
| Pisces → Stegocephalia | |
| | |
| Origin of bird from a reptile's egg? | |
| Two periods of differentiation of land plants | Fig. 18 |

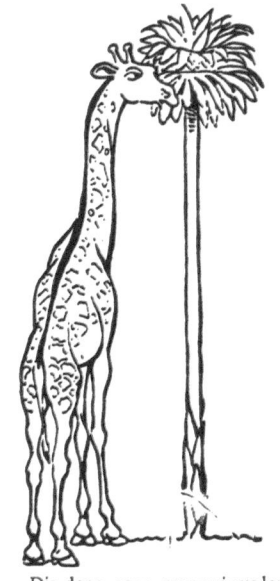

— Dis donc, papa, *pourquoi* que les palmiers sont si grands?
— C'est pour que les girafes puissent les manger, mon enfant, car...

...si les palmiers étaient tout petits, les girafes seraient tres embarrassées.

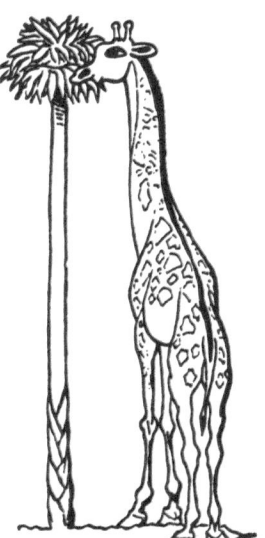

— Mais alors, papa, *pourquoi* que les girafes ont le cou si long?
— Eh bien! C'est pour pouvoir manger les palmiers, mon enfant, car...

...si les girafes avaient le cou court, elles seraient encore bien plus embarrassées.

Fig. 115. Lamarckism in caricature, by Caran d'Ache. (After Sirks).

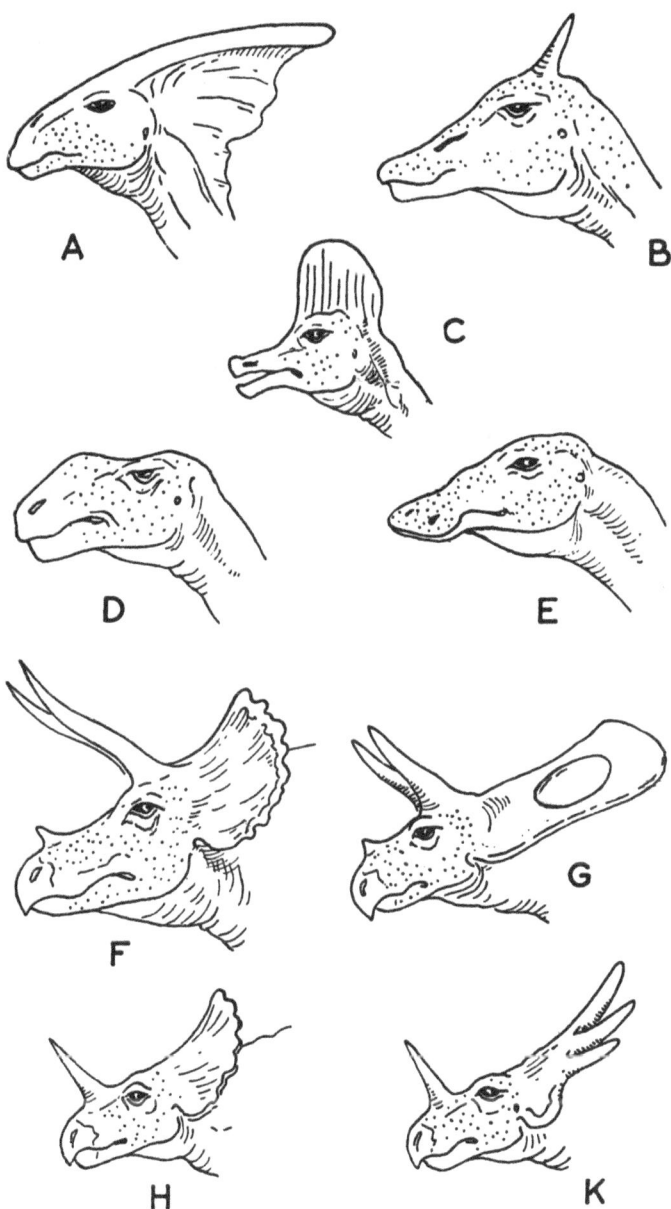

Fig. 116. Some of the last Dinosaurs of Upper Cretaceous times, A, Parasaurolophus; B, Saurolophus; C, Corythosaurus; D, Gryposaurus; E, Trachodon; F, Triceratops; G, Torosaurus; H, Monoclonius; K, Styracosaurus. (Adapted from R. S. Lull).

organism to its organic and inorganic environment by the inheritance of characters acquired during individual development (see fig. 115).

It goes without saying that the idea of invention leads us to invest the directing vital principle with qualities which we know from our own mental activity. Thus this purely anthropomorphic principle (sometimes designated by a name such as *entelechy*) is looked upon as if it were an artist or an architect who has taken pleasure in reproducing beautiful and complicated organs, instincts and organisms. Endowed with human qualities as this life-force is thought to be, it also has its weaknesses. Sometimes it has a fancy to imitate, or to equip strange-looking creatures with organs or characteristics which exceed the bounds of utility (fig. 110). Then again it finds its choice in the composition of its products restricted by the nature of the material that it has seized upon or the influence of the environment in which the new creature is to be let loose. But does that make any difference? Even Jupiter and his fellow dwellers on Olympus were not entirely free, gods though they were. At another time this vital principle enjoys the devilish satisfaction of annihilating with one blow one of its own products that has been steadily developing during some tens of millions of years. Did this creative spirit, after having watched the evolution of the Dinosaurs for nearly a hundred million years, all of a sudden tire of them and wipe them out of existence with one sweep? (fig. 116). The vagaries of this regulating, creating and inventing vital urge can, moreover, be put to the test by biological experiments.

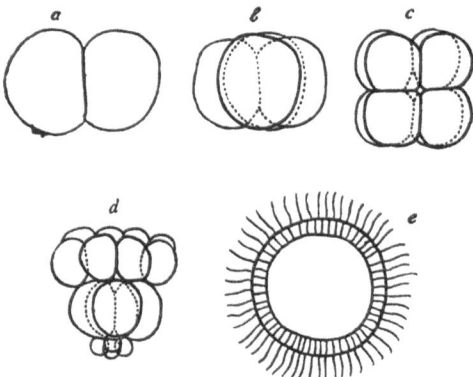

If the development of an organism were merely a matter of physico-chemical processes, which run down mechanically like a wound up watch, then it is logical to suppose that a fertilized egg which has been cut in half cannot possibly produce a complete organism. With eagerness, therefore, Driesch in 1891 looked forward to the result of his famous experiment with the first stages of segmentation of the eggs of sea-urchins. After fertilization the egg splits up into 2, 4, 8 etc. parts, called blastomeres (fig. 117). The first two segments were separated from one another along a vertical

Fig. 117. First stages of segmentation of the egg of a sea-urchin (Echinus) a–d, the egg splits up in 2, 4, 8, 16 parts; e, blastula stage (After Driesch).

plane. The result of the further development was not two halves of a sea-urchin, but two complete ones. When repeated after the second, third and fourth cell divisions, the experiment gave the same result. It appeared, therefore, that it was not the fertilized egg as a whole that had the power of producing a complete individual, but small parts of it. Then, Driesch pressed the blastomeres together between two glass plates, thereby altering the direction of the segmentation. When left to themselves there came not a morphological chaos, but again a normal individual (fig. 118). The obvious conclusion is, first that the development does not run down like a machine, and is therefore not preformed, and secondly, that there must be a wilful, non-

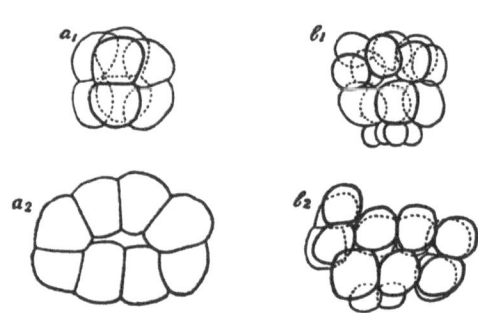

Fig. 118. $a_1$ and $b_1$ normally developed blastomeres of Echinus; $a_2$ and $b_2$ the same after having been compressed until reaching the stage of 8 blastomeres. (After Driesch).

material force inherent in life, which is capable of bringing the ontogenetic processes to a normal conclusion, even — as Professor L. Rutten said — :"when biologists have teased and maltreated the poor egg in a manner which no dog could endure". This directing vital principle Driesch called *entelechy*, a term taken from Aristotle, but used with a different meaning from the original one.

Entelechy seems to be the force which creates order in the organism. It seems to be so, but further experiments have made it doubtful whether this supposed immaterial, regulating intelligence exists. An isolated blastomere of an Ascidia or Ctenophore does not yield a complete individual; and, before Driesch, Roux had already killed the half of a frog's egg after the first cell-division, with the result that only the half of an embryo was produced. Furthermore, when the segmentation stages of the sea-urchin are not halved along a vertical but along a horizontal plane a complete individual is not formed. Moreover, Driesch's experiments, and therefore also the conclusions drawn from them, only succeed with very young stages of segmentation. In a more advanced stage the experiment fails, and one may well ask: has the powerful entelechy then withdrawn from the individual, leaving it to its material fate, just as, according to Driesch, it also leaves the body when death supervenes? Finally, if the entelechy is so intelligent as to secure that a mutilated starfish regains its original five-armed shape, why is this same intelligence so stupid or capricious as not to heal certain injuries to a Planaria normally, but to make of it a monstrosity with many heads and tails (fig. 119). All this can only mean that the theory, at first so obvious, of an entelechy as an immaterial life-principle, operating beside and distinct from the physicochemical processes in the organism, on further investigation turns out to be a hypothesis which breaks down entirely on many points, even when one is prepared like Whitehead (as will appear shortly) to allow for the devil as well. It would not be difficult to spend much more time in expatiating on such qualities of a vital force, supported by countless examples from the animal and vegetable kingdoms and the lessons taught by their fossil remains.

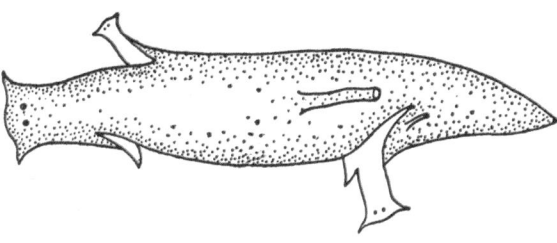

Fig. 119. Monstrous regeneration of a flatworm (Planaria). (After Wells and Huxley).

Now it is easy enough to show the fallacy of these so-called finalistic or teleological attempts to explain the characteristics of life. But it is far more difficult to replace the stultified image of final causes by something positive. When we describe the characteristic of a living creature as an inherent influence which is continually exerted upon the coordinated whole of the organism (as contrasted with e.g. a beam of α-particles or photons), or again when we aver that it shows signs of an inherent efficiency (= usefulness), all such descriptions imply a constituent principle that is definitely directed. We are then immediately inclined again to have recourse to the word "purpose". Why should we not also speak of a purpose (as has indeed been done by W. R. Thompson in his "Science and Common Sense") in the case of a pure direction of the phenomena, e.g. of a beam of α-particles directed towards the north? It is only a question of definititon or agreement, and there is no objection at all (except that it is confusing to use the same term) to speaking of the finality of life, provided one does not attach to it the anthropomorphic background of the artist or inventor with personal human qualities who made a pre-meditated plan or had the intention of achieving a certain effect.

The following may serve as an illustration. Take ten marbles and let them all roll together from your hand on to the floor. Where each one comes to rest is a matter of chance. It is different if the ten marbles are one by one rolled in a particular direction, e.g. towards the southwest, without any particular resting-pont being intended. But thirdly it is possible not only to direct the marbles in a particular direction, but at the same time to have the intention of shooting them into a clearly seen hole so that they shall come to rest in that hole and nowhere else. In the second and third cases both "intention" and "finality" could be used in a different sense. In other words, direction may be accompanied by a clearly formed purpose as to the result, but this need not a priori be the case.

It has just been argued that the development of a living creature and the phenomenon of evolution cannot be compared to the third instance of the marbles. That they can also not be compared to either of the other instances will be discussed later in this chapter (p. 191).

The great difficulty is that no human conception, expressed in an intelligible picture, can represent the characteristic of life.

These difficulties, the limited powers both of conception and of expression in suitable terms, are clearly illustrated by the valuable speculations of Whitehead. There is the more reason to discuss these here, as Whitehead's attempt to define life is based on characteristic qualities of man himself.

Whitehead's tendency to take man's qualities as a reference for the understanding of the whole of nature leads, it seems to me, also to the idea that the imperfect, the unnecessary, and the faulty, whichever one wishes to call it, in nature (especially in the phenomena which biology and evolution reveal to us) must be regarded as the necessary, and therefore comprehensible opposites of the perfect, the desired and the useful which is to be found beside it. In other words, he too endows the vital principle with the same virtues and vices, and the same well-meaning but fallible qualities of a human being; and the problem seems to be solved, provided that we do not look for perfection, but try to distinguish the perfect and free, and define it as the counterpart of the fallible, imperfect and unfree. It seems clear to me that in these latest theories of Whitehead, too, an image is fashioned; an image that is no other than that of man himself — and his problems.

The basic idea which in various forms is continually to be discerned in Whitehead's work comes more or less to this. When we ourselves experience a sensation, something more happens than a mere passive reaction to perceptions or impressions. Every experience contains at the same time the expression of the result of reflection. The former regards the causa connection with previous experiences, or the past; the latter opens up a view of the future and contains a consciousness of value, perhaps even of responsibility.

On this wholly anthropomorphic basis is founded, in the interpretation given to it by Burgers, the theory "that the basic form of all nature is a manifold and infinite system of interrelated events, each one of which originates from a particular experience of existing facts, and forms the materialization of a correlation which has been perceived and assessed in that experience, while by this materialization a new fact is immediately formed, which is added to the rest, and thereby becomes with them the basis for new events".

Experience and value, however, are concepts which are most closely associated with a human and conscious power of discrimination. It is very much to be doubted whether we are justified in making something typically human like this the elementary principle on which the whole of living beings is based. Consequently, the term value, placed between inverted commas, should also include the reason "why something reacts to an experience as it in fact does".

This means that the term is indeed retained, but not its anthropomorphic significance, when we are dealing with the description of processes in other organisms — and *a fortiori* of inorganic processes.

What then remains as the common characteristic of all processes is, according to Burgers, the discontinuity, the repeated succession of synthetic pauses, which contain the "creative" factor for a new action and possess an element of spontaneity and freedom. Here again we have a series of descriptive terms which have a strong anthropomorphic tint, and are consciously taken from what we find in ourselves.

Spontaneity and freedom are regarded as the basic principles of all processes in the whole of nature. In inorganic processes these elements of spontaneity are said to compensate, and so suppress one another. On the other hand Burgers, following Whitehead, considers it possible that perhaps "special linkings" will lead to a strengthening of an aroused sense of value and thereby also reveal the spontaneity in more complicated reactions. The concept "life" then requires the repeated occurrence of such more complicated spontaneous reactions, which must at the same time be so directed that they do not endanger the continuance of the group of processes within which they have appeared. This means that the basis of life, spontaneity, is regarded as something that belongs to the first principles of all nature; the complicated manifestations of it which we see in living organisms are thought only to be brought about *by a special ordination, the meaning and origin of which we do not yet understand* [1]).

Put very shortly, it comes to this, that Burgers, in his above mentioned interpretation of Whitehead, uses purely anthropomorphic concepts or images as his starting-point, but, in applying these concepts to the organisms and inorganic processes which surround us, cannot maintain the anthropomorphic significance of the terms he uses, and goes on to consider the difference in the natures of living and non-living systems to be only incidental; so that finally a large question mark must be attached to the origin and significance of these hypothetical differences. But this is the same thing as confessing that the vital urge, which characterizes both our lives and the evolution of the whole world of organisms, is governed by a principle which does not imply either perfection or forethought, but is unknown, incomprehensible, and incapable of being visualized. It is a principle in which perfection, freedom etc. and their opposites are no more than certain attributes of life which are to be found only in ourselves, and are consequently manifestations which are only to be visualized with reference to human beings.

It is, therefore, wrong to use the word "creations", to cover at one and the same time, as if analogous to one another, the "creations" of nature and those of an artist. It is wrong for two reasons. First, because to include a work of art and the "creations" of nature in one synthesis implies that it is possible to answer the problem of the origin of the universe and its parts (i.e. the "creation" or conception of galaxies and flowers, kangaroos and electrons, dinosaurs and gnats, of "life" in general and human life in particular) by means of the same pictorial imagination which enables us to understand the work of an artist. Rightly did Alexander therefore say:

"But when we ask ourselves about the creation of the world, we stumble, "because we carry these pictures about with us and we try to interpret by "their help what is beyond the reach of pictures".

In the very presentation of the problem an element has been introduced which makes it impossible to give an answer which is free from an inherent contradiction. The second error consists in the premise of a creative intelligence outside nature, which may be compared to the creative intelligence of the artist outside the material on which he is going to work. That is what one does when nature is regarded as *artis magistra*. In this manner did Michael Angelo depict his creation scenes in the Sixtine Chapel. In this manner, too, nature was formerly without exception, and is still by many, regarded as the product of an intelligence which we

---

[1]) Translated from Burgers, 1942, p. 141 (the italics are mine).

picture to ourselves as analogous to the intelligence of an artist or an engineer.

It is fortunate for him who reasons thus that he stops there, and does not feel any desire to go logically further with an enquire into the origin of that creative intelligence and the fiction of infinity which is bound up with it for, he would then only become tied up in the inevitable, and yet inconceivable, consequence of a beginning and an end, the logical and at the same time absurd conclusion that something must have come from nothing, and so forth; without, however, realizing that these insoluble problems arise from the initial hypothesis, his own hypothetical image, which thus leads to only one indisputable conclusion, namely that it is in contradiction with itself.

Nor can he see that the opposites unite in one whole, when they are stripped of the artistic imagery (see p. 211–213).

*Materialism.*

Pre-meditated purpose is an entirely superfluous concept in the system built up on the theory of selection. For, according to the Darwinian view highly efficient organs can be produced by the blind play of chance or, to put it in another way, as the inevitable result of a series of physical and chemical processes, whose number and interrelation we can as yet only imperfectly describe. Terms such as finality and teleology are therefore not to be found in the vocabulary of the theory of selection. Teleology is a concept produced by the psychology of human society, where it has a real meaning. A human being can point out to himself or others a goal towards which he can direct his conscious intention. But consciousness is a factor which cannot have played a part of any significance in evolution before the appearance of human beings. Darwin therefore rightly held teleology to be a purely anthropomorphic simile without any value in explaining the phenomena of life.

Teleology is a concept which is already to be found in Anaxagoras and Plato, but in the Middle Ages, particularly owing to the influence of Aristotle, it dominated the interpretation of nature. The work of Galileo, Descartes and Newton expelled it from the realm of physics. Finally, two centuries later, biology was also released from this obsession of intentional vital activity as a leading principle. When, therefore, some of the neo-vitalists, as we have seen, while rejecting the theory of selection, reintroduced the concept of final causes by another door, they went a step backwards instead of forwards along the arduous road of the search after the secret of life.

In Darwin's doctrine the idea of final causation was repudiated and replaced by the theory of natural selection. As a breeder is able by crossing and subsequently selecting variations to breed an astonishing diversity of domesticated pigeons from the wild rock pigeon which formed the material with which he started, so it was thought that natural selection was responsible for the differences between the various organisms. In the struggle for existence only those of the many arbitrary variations will survive which are useful to the organism. Thus we get not only a diversity of organisms, but also a progressive adaptation to different environments. Chains of cause and effect take the place of teleology, physical causation that of final ends.

Now it is impossible to explain the origin of diversity in the organisms otherwise than by presuming that certain random variations in individuals, which are thereby better equipped than others, by continual repetition in countless successive generations constantly develop along the lines of the greatest usefulness, while others, which do not fit the final product as we know it, have perished and been eliminated. The same thing, in fact, as happens in the case of artificial selection by a breeder (apart from the condition that the useful variations in nature must also be hereditary). Stronger still, paleontology has given us numerous examples of a pheno- menon which is called orthogenetic series: a definite direction in the variational changes in characteristics or organs, covering sometimes many tens of millions of years. This has convinced many scientists that this phenomenon cannot possibly be ascribed to the play of fortuity.

The following example has often been given to show what is meant by fortuity. When a tile falls from a roof and comes down exactly on the head of a very important passer-by, it is called fortuity. But, as a matter of fact, the accident necessarily happened thus, the circumstances being what they were. It is said to be chance because the result is caused by a number of untraceable and insignificant factors; this is perhaps still clearer when we think of a roulette or any other gambling game. Chance or fortuity is the name given to the striking coincidence or concourse of two or more independently occurring events, each of which in itself forms an undeniable chain of cause to effect.

When, however, tiles continually fall from the left and the right on to that one person, we can no longer believe that it is only fortuity. The projectiles are definitely directed in too long a series for that; they are being aimed at him. The history of life shows us thousands and thousands of definitely pointed coincidences in continuous series. Such phenomena cannot, therefore, be caused merely by chance. Let us remain as neutral as possible in our nomenclature, and speak of a definite direction or trend of the events.

Some authors hold that selection is a directive factor which under certain circumstances causes a phenomenon like orthogenesis. They even speak of orthoselection.

The following argument is, however, frequently put forward by opponents of the selection theory. Granted that certain characters, e.g. the antlers of a stag, have selective value, why will an increase in their size be favoured by selection when they are barely incipient? Another example quoted from Simpson's book reads as follows: "A complex ammonite suture has a conceivable advantage over a simple suture, but in the gradual change is a barely perceptible increase in complication of sufficient value to make selection favour continued change in that direction? Some degree of mimicry is evidently adaptively useful, but do predators really distinguish between approximate resemblance and the very exact resemblance often involved in mimicry? Such examples, which are very numerous, are standard items of evidence in favour of an inherent factor of some sort in orthogenesis, independent of selection".

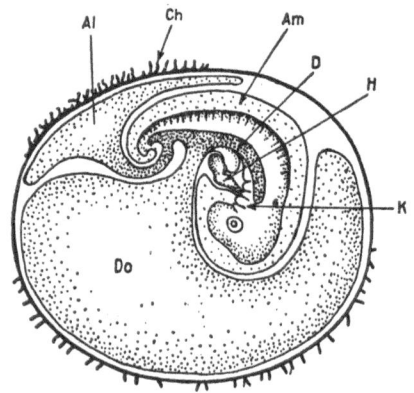

Fig. 120. Embryo of a rabbit, showing coenogenetic organs. Al, alantois; Am, amnion; Ch, chorionic villi which will form the placenta, D, intestine; Do, yolk sack; H, heart; K, gill region. (After Van Beneden and Julin, from Ziegler).

Grave difficulties are also offered by phenomena of metamorphosis (see p. 193, 194, and fig. 122) and by so-called coenogenetic organs. During the development of some organisms organs come into being which are only useful for the embryo at a certain stage of its development. They disappear at a later stage of development of the creature. This embryonic specialization is termed *coenogenesis*. An example is the yolk sac, which serves as nourishment for the embryo. A similar feature is the seed-lobe or cotyledon in a germinating plant. Other examples are organs like the allantois and placenta in mammals (fig. 120).

How orthogenesis must be understood according to the theory of selection is explained by Julian Huxley more or less as follows.

At all times there are on the one hand in the environment allotted by nature open spaces which are waiting to be occupied; on the other hand many improvements can still be made in existing organic types. These

two possibilities overlap each other. When, to give an example, the power
of perception of an organism undergoes an improvement, the creature
will also be able to conquer new room to live in. Changes in the structure
of an organism may occur at random, but those which prove to be an
advantage in the struggle for life are preserved and increase the maximum
possibilities of life; and this goes on and on as the result of the constant
occurring of new fortuitous variations. There is, as it were, a premium
attached to the occurrence of a progressive variation, whether it is a small
variation in a part of an organ or a saltation of greater extent. This is the
process, the so-called "urge of life", which is called orthogenesis. But
this does not mean that there is a mysterious *élan vital*; it is merely a
question of random variations, which, however, continue to tend in one
particular direction, for the reason that it is only in that direction that
they are of ever increasing importance to the organism in its struggle for
existence, and their selective value therefore surpasses the average of a
hitherto existing level. The struggle for existence and the elimination of
the less fit by natural selection therefore works in such a manner that of
everything new provided by chance only the valuable, the improved equipment is retained. Selection is therefore like a filter, a sifting agency which
guides the random variations in a direction prescribed by the particular
conditions of the environment. Just as, for example, the invention of a
motor car with four cylinders had a greater "survival value" than one
with two cylinders, and so supplanted it; only to be subsequently in its
turn superseded when the "orthogenesis" progressed still further and
motor cars came on the market with the selective advantage of six or eight
cylinders. Yet there are still circumstances in which a car of the old type
is useful. So there are also spheres of life in which the organisms which
are not so strongly specialized and less adapted to particular circumstances
can find means of existence; so that we still find types of the original stock
persisting side by side with highly evolved types. While on the one hand
these primitive organisms continue to exist, on the other hand there is a
slow but sure increase in the complexity of structure, an advance of the
highest level of the flow of life, manifested in ever better equipped types
of life. In this progressive stream we frequently see a long continued
increase in size, often accompanied by greater longevity of the individual;
further a growing functional differentiation in the parts of the organism;
then a greater harmony of the parts (coordination, unity of the whole),
and an increasing self-adaptiveness of the creature to the continually
changing circumstances of his environment. In the most advanced stage
there comes the capacity to apply experiences of the past to problems of

the present. Further there awakes what we know as psychic specialization; human knowledge, emotions, conscience and free-will begin to play a part in the life of the individual and the community. We have arrived at the highest product of the mechanical workings of variations and selection — Homo sapiens.

The reader who has followed this argument carefully will remark that the filtering action of selection can only take place if there is a continual occurrence of random variations of a very special sort and in combination with other correlated characters. The objection therefore remains that the theory asks too much of us when it will have us believe that directed or straight-line evolution on a scale as shown by paleontology could be determined by chance (random mutation) combined with natural selection.

To understand this the assistance of another theory (the so-called *autogenesis*) is given by some authors. The fact that the same characteristic continues to appear, sometimes even in an increasing degree, is said to be the result of certain internal physical and chemical processes in the germ-cells not involving natural selection ("ectogenesis"), which restrict the potentialities of the germ plasm to very narrowly limited possibilities. Eimer has already sought to solve the problem in this direction. To give an exact definition of what these processes are may be a problem for the future, but in no case is it necessary to resort to a mysterious *élan vital* or any other irrational principle in order to explain the urge of life. This theory of the so-called *autogenesis* is said, moreover, to help us out of the difficulty encountered when dealing with such a phenomenon as homomorphy which raises still more obstacles in the way of the theory of selection than orthogenesis does. But this and many other difficulties cannot be dealt with here, seeing that we are here concerned only with following a main line or general principle. We will return to the theory of autogenesis in the next section.

Some scientists have rejected the selection theory entirely. Yet, it would be absurd to dismiss the idea that selection eliminated at least the inadaptive types of organisms. At any rate we must consider it as a conservative force, which as it were tries to stabilize the position of the population curve. However, according to modern theories of population genetics, selection is a truly creative though complex process acting on changes in gene frequencies in populations produced by differential reproduction. It is therefore held that under certain circumstances selection tends either to shift the position of the curve in a special direction (cf. fig. 96) or to split it into two or more divergent directions (cf. fig. 94).

In this respect the researches of geneticists are of great importance and we may confidently await important results pertaining to the role played by selection in the intricate processes of evolution. Several authors have probably over-estimated the importance of selection inasmuch as they considered it as a sacrosanct principle which was thought to explain all phenomena in the evolution of life. This attitude inevitably leads to a materialistic discipline concerning the problems of life and the world in general, to a purely "causal" description of the phenomena of life and its evolution in particular.

In this way the succesful results obtained by nurserymen, breeders, and experimental biologists led Darwinists to the conviction that both the physiological processes of a living organism and the evolution of the vegetable and animal kingdoms can be explained completely and solely by concepts of chemistry and physics (including the part played by chance). The origin of heavenly bodies from primordial chaos, the differentiation of the chemical elements and substances, the emergence of organic out of inorganic matter, the grouping of organic matter in ever more complex protein molecules, and from these the formation of the most primitive conglomerations which can be called alive; and so, by increasing differentiation on the same lines, from the original cell to the high level of that product of blind chance — Homo sapiens.

Thus grew the materialistic monism of Haeckel. which is characteristic of the end of the nineteenth century. As to the problem of life, selectionists and Darwinists can even now regard it only in‹this way: there are no problems which cannot be unravelled by means of the exact methods of chemistry and physics with which we are familiar; none other than a "causal" explanation for life is accepted; and all arguments tending in any other direction are a priori stigmatized as unscientific and metaphysical and are rejected without further consideration. Evolution is merely a question of selection or it is a combination of selection and random mutations, or it is a purely automatic result of "causally connected" physico-chemical processes in the germ-cell (according to the so-called autogenesis theory), or it is a combination of both; but in this conception of the world there is no room for factors which do not belong to the realm of physics and chemistry. It may be granted that at present we still have to attribute a great deal to chance, but as the different physico-chemical processes involved become more accurately known, we shall be able to replace the factor of chance by ever better understood chains of blindly inevitable cause and effect. Such was the picture of life conceived by the positivist monism of the nineteenth century, in essence resembling a chemical factory, which is either functioning automatically or which is directed by the interplay of random mutations and selection.

When a theory of such consequence has the charm of novelty and is propagated by such ardent crusaders as Thomas Henry Huxley and Ernst Haeckel, and when it has the support, as is the case even today, of so well-known a biologist as Julian Huxley many are gladly willing to regard its weak points as of relative insignificance, assuming that they see them at all. Thus many a neo-positivist is even now satisfied with this picture of life [1]). To them it seems as if the time is not far distant when everything will be known. In outline there is a complete, logical picture of the world, for all time as it seems. The main conception of the structure needs no further alteration; there are only a few gaps here and there to be filled, a few details to be embellished. The scientific investigator of the future can hardly be thought of otherwise than as a complacent gentleman, twiddling his thumbs beside a retort with chemicals from which, according to his calculations, after exactly 43 minutes and 12 seconds a human foetus will emerge, crowing with vitality.

*The concept of holism.*

However, in the words of Jeans: "We are beginning to see that man had freed himself from the anthropomorphic error of imagining that the workings of nature could be compared to those of his own whims and caprices, only to fall headlong into the second anthropomorphic error of imagining that they could be compared to the workings of his own muscles and sinews". For the basic insufficiency of a "causal" description of the phenomena of life has already been clearly revealed by many scientists. Insufficient — for although the analyst is right when he shows that e.g. the physiological processes in an organism follow the laws of physics and chemistry, he forgets that an organism does not consist of a loose conglomeration of causal chains, but is an individual whole which contains more than the simple sum of its parts. So also the "geological history" of organisms shows us that there exists linearity in evolution not only of one or more characteristics, but of a coordinated whole, for which mechanics and chance cannot give an adequate explanation.

This may be made clearer with the aid of one example, of which even

---

[1]) Next to the extreme view which considers the explanation of the origin of new types and more complicated systematic entities to lie in selection alone, coupled perhaps with *autogenesis*, other biologists prefer to attach less importance to the part played by selection. Others again attribute only a conservative, stabilizing function to natural selection. The scope of this chapter, however, precludes going into this further.

Darwin admitted that it was a nightmare to him to think of it — an example which Huxley in his above-mentioned defence of Darwinism passes by as if there were no difficulty attached to it — namely the development and evolution of the eye. How can we regard our eye as the product of a fortuitous combination of a large number of parts which together form an efficient organ — the lens, cornea, iris, and aqueous and vitreous substance, the muscles and nerves connected with it, the power of adjustment, etc.? It goes with a certain type of nervous system and forms a complete whole appropriate to the organism. During the formation of the eye a number of processes follow one another, and matter is conveyed to it from adjacent or more remote parts of the body in such a complicated sequence and mutual relation, that we can only speak of a coordinated interaction of events which, operating in due proportion to one another, lead to a harmonious whole. The eye of a Nautilus or a Tridacna is of a much simpler construction, but it is an organ which is suited to those organisms and functions with natural efficiency. It is absolutely impossible to imagine that a mechanism such as the human eye could, in the course of evolution, be derived from such a primitive eye merely by the addition of a series of fortuitous variations with chance to be of benefit to the creature.

The physiologist H. J. Jordan sums it up as follows: "The formation of new organs from abnormalities which are so small that their occurrence may be considered to be fortuitous variations and yet are of use to the creature, is an entirely different problem, for which experience has not yet been able to supply a solution. Examples can show that Darwin's theory does not explain it at all. In the lower classes of animal life we find eyes that do possess a lens, but are not able, when the distance from the object is changed, to focus the lens in such a manner that the picture still falls on the retina. On the other hand we know of animals, human beings for example, with whom this so-called adjustment does take place. As contrasted with the former sort of eye the adjusting eye (e.g. of a human being) requires the following peculiarities (see fig. 121). Not an inflexible, but an *elastic lens* [1]), contained in a *lens-capsule*. This lens-capsule is connected with the surface of the eyeball by the *taut threads of the Zonula zinnii*, so that the tension makes the naturally convex lens flat (adjustment for distant view). A *muscle* on the surface of the eyeball, the adjusting muscle, by contracting relaxes the lens-capsule, so that the

---

[1]) In this quotation italics have been used for the most important factors, which would all have to come about together in order to turn an inadjustable into an adjustable eye.

elastic lens can become more and more round. Thus the lens is enabled also to concentrate the rays on the retina when the object is nearer (adjustment for near view). This, however, requires a reaction of the muscle to the different distances of the object viewed. The retina communicates this distance through the optic nerve. This must be connected in a *part of the brain* with the root of the adjustment nerve (which leads to the

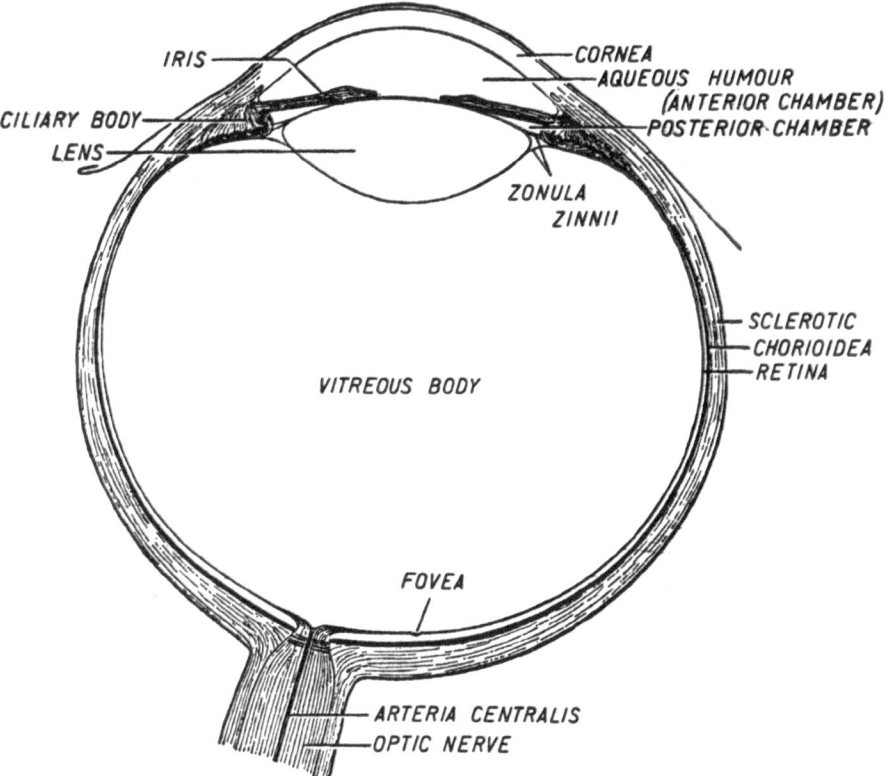

Fig. 121. Anatomy of the human eye.

adjustment muscle), and the action of the muscle must be exactly *adjusted* to the distance of the object viewed. Not only must all these parts be present, but they must also be present in a fixed mutual relation, and the extent of the functioning of each must be in precise concordance with that of each of the others. Each single one of this series of factors alone is pointless, and even injurious. A snail with an elastic lens, which is not controlled by an adjusting mechanism, sees less than an ordinary snail, and will certainly not survive in preference to the other snails to become the first ancestor of animals with adjusting eyes. The problem is a universal one. There is no organization which is not a compuond of numerous factors, of which none can be dispensed with and which must all be present in a fixed relative order and harmony — the absence of any one meaning inferiority. Therefore, the whole cannot be only the result of the mere addition of the parts. The parts of every organism have at least an "amboceptor" character, that is to say that they always fit in with at least two adjacent factors, so that all the factors together form a network of parts, whose qualities are exactly coordinated with one another. In such a network it is impossible to assume that new elements can be introduced by chance" In the words of H. J. Jordan:

"Organs with specific functions imply the existence of the appropriate instincts; nothing is conceivable by itself alone; each part separately has no reason. Mutations must have been the cause of new phylogenetic stages occurring, but these mutations must have produced a plurality of factors in a fixed relative order; factors, therefore, with an amboceptor character ("amboceptor mutations"). Such "amboceptor mutations" are purely hypothetical. They can only have come about through a plurality of causes in a fixed mutual relation (pluricausal origin with immanent order). The attempt to explain the origin of the different organisms by selection coupled with fortuitous (monocausal) mutations cannot be considered to have any scientific value".

This seems so in spite of the fact (which is the last argument of materialism) that in analysing the organism one continually perceives the material processes and their separate chains of causation, and in spite of the fact that experimental embryology has already solved many details of the working of certain substances (so-called organizers and inductors); and, finally, in spite of the admission that we are still entirely in the dark concerning the nature, the working and the meaning of the coordinated relation of the separate chains of causation.

For when we speak of pluricausality, laws of linking, causality of composite structures, or of ambo-, poly-, and pantoceptor relations, all these words throw no light at all on the problem of the composite harmony to which they refer. That is more or less the point of view which has drawn the attention of many biologists in the last fifteen years since Smuts clearly defined it and, under the name of *holism*, set it up against materialism and vitalism.

Indeed, how can the metamorphosis of a caterpillar into a butterfly, the life of a Balanus or a parasite such as the Sacculina (fig. 122), the life cycle of a malaria mosquito or a Trypanosoma, the actions of a leafcutter bee, and a thousand similar examples be comprehended in the light of the same laws which suffice to describe the formation of a diamond or a quartz crystal, the movements of an aeroplane or a pendulum clock, and the chemical action of saliva or gastric juice? It is true that an analytical investigation of growth and the mechanism of movement and all physiological processes reveal no other laws than those with which we are familiar in mechanics, statistics, chemistry, and physiology. But the main problem of biology is not that the movements of a dragonfly and a

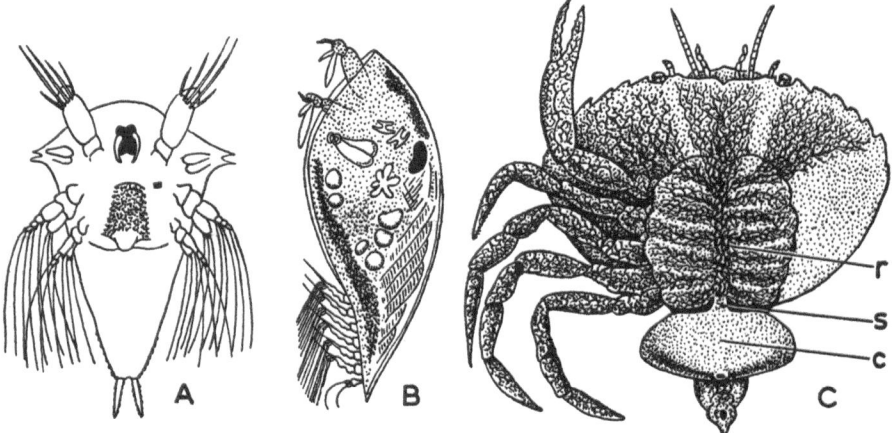

Fig. 122. A and B two successive young stages in the development of Sacculina carcini; A, Nauplius stage; B, Cypris stage; C, adult stage. The organism is deformed into a sac-like parasite (c) attached to the underside of the tail of a crab. By way of the stalk (s) the sack is connected with the ramifications (r) in the body of the crab. (Adapted from G. Plate).

caterpillar somewhat resemble those of an aeroplane and a tank, and that acids and alkalis in the body act in the same manner as acids and alkalis in a test-tube, but that a dragon-fly and a caterpillar are functioning organisms in which the laws of matter have been organized into one coordinated whole for the maintenance, growth, protection, and pro-creation of the creature.

The concept of holism has been applied to organic evolution as a whole by Leconte du Nouy in his book "Human destiny". The urge of life displayed during the last 1200 millions of years is compared by him with the "urge" of water which running down from a mountain-slope under the force of gravity will always try to flow towards a certain predestined goal, sea-level. The water may divide into a number of streams, some will grow larger, others will lose themselves in marshes or in rock fissures. Some will end in lakes and they will encounter countless obstacles on their way, which will deflect their course or mould their shapes.

In a similar way the evolution of living beings is visualized as a whole which tends to reach a certain level, a predestined goal. In doing so evolution should be considered as an irreversible progressive phenomenon. Chance and changing physical conditions of the environment will cause local and temporary deflections of the evolutive impetus, they will stimulate the occurrence of sudden mutations, they will give origin to a host of adaptations and specialized types. And these are doomed as soon as the conditions to which they were so excellently adapted change. Other branches will continue blindly and end in troublesome or even harmful monstrosities (fig. 110 and 116). However, these are nothing but negligible accidents. They represent "the tail end of a strain long before separated from the evolving stem". What matters is the fate of the species "considered as a link in evolution as a whole".

This is the holistic theory of Leconte du Nouy, which he indicated by the term *telefinalism*. He rejects the old-fashioned finalistic theories, and regards telefinalism, which would dominate evolution as an essentially different process, a sort of evolutive impetus which orients the march of evolution as a whole. According to the concept of telefinalism there is a presupposed goal which "ever since the appearance of life on earth tended to develop a being endowed with a conscience, spiritually and morally perfect being". All other manifestations of life in our present fauna and flora as well as during the 1200 million years of organic development are considered by him to represent only the "left overs" of evolution. Man is considered as the only exception!

Specialized creatures belong to side lines. They will disappear from the stage sooner or later. Only the non-specialized unstable forms are capable of still greater changes, of evolution. In Leconte du Nouy's opinion "only one strain amongst all the others never attained equilibrium and yet survived. This was the line that ended in man". "Evolution continues. It continues through man and through him alone". Similar views were expressed by Vandel, though he starts from other, biological considerations and instead of telefinalism speaks of psychism and a universal intellect. As to the privileged position of mankind Huxley, too, wrote: "Finally, but one line was left which was able to achieve further progress; all the others had led up blind alleys. This was the line leading to the evolution of the human brain". Not every paleontologist will agree with this point of view. Man is mostly considered as a specialized creature, a specialist in mental capacities. "Regrettable though it may be", Hawkins wrote, "the human animal seems to a palaeontologist superior to a Dinosaur or an Ammonite merely in the speed with which it rushes towards extinction". Organisms which evolve under favourable circumstances, often bear in themselves the germs of their total extinction. For also the unfit and pathological individuals have a great chance to survive and to take part in the procreation of the species. In the long run the effect of his mental specialization may turn out to be the weakest point of humanity. A few specialized organisms like Terebratula and Nautilus representing types which persisted for more than 250 millions of years, show that there are exceptions to the rule expressed in Hawkins' verdict. But even if human beings were to disappear from the stage would this mean the end of evolution? Are there no unspecialized or slightly specialized organisms capable of further evolution? Nobody can foresee which group might evolve in such a way that it would become pre-destined to take over man's supremacy of the world as well as, though possibly not necessarily, to outstrip Homo sapiens effectively in the capacity and quality of his brains.

Man is different from e.g. ants, bees, and termites; however, saying that man has evolved higher than these creatures would be an assertion without sense. For the evolution of arthropods cannot be measured or evaluated in terms of vertebrate evolution. Even if all vertebrates, including man, became extinct would that mean the end of evolution for all other animals and plants?

We shall not enter into the further aspects of the remarkable book by Leconte du Nouy. Suffice it to say that his further arguments lead him to postulate the development of Christian ethics as the main goal of humanity. According to his opinion an interpretation of the organic world without adopting the telefinalistic point of view completely escapes our understanding. However, telefinalism postulates "the intervention of an Idea, a Will, a supreme Intelligence". This means the introduction of an anthropomorphic principle, which on du Nouy's own advice we must carefully avoid and which, moreover, explains neither the meaning of the whole play of evolution nor the origin and meaning of the principle which it postulates. It merely expresses the religious conviction of du Nouy, in the same manner as the writings of Aristotle in the interpretation given by Thomas Aquinas is the only sacrosanct doctrine to others.

The finalistic postulate in so many theories is always invoked because it "seems to be the only one which gives man a reason for existence and attributes a definite significance to his life". Substracting the anthropomorphic ideas from the telefinalistic hypothesis means frankly admitting that we cannot possibly interpret the evolutive impetus in terms of an imaginative picture borrowed from our human experience and surroundings.

This, surely, will sound too defeatistic to many people because of the trouble they meet in reconciling it with their ethics or religious convictions. However, even if Homo sapiens were not the goal of evolution he may, as a specialist in psychical capacities, continue proudly to stand at the present top of the ladder of systematic classification and vertebrate evolution. Moreover, he has developed moral standards and a conscience. More than that man is able to set himself a goal and, by all means, he may try to propagate among his fellow brothers the value of human dignity — and behave himself accordingly.

*Summary of conclusions.*

The above review has yielded some "negative" results of importance. On the one hand the failure of the materialistic theories shows that a living organism can *not* be completely described by means of physico-chemical processes alone. On the other hand the unsuccessful attempts of the neo-vitalistics and finalistic theories demonstrate that life can *not* be regarded as a separate principle which, itself independent of physico-chemical processes, orders these processes and coordinates them.

The characteristic of life (to put it shortly and therefore inadequately) consists of a principle which is bound up with the material processes of the organism into harmony, and makes the organism react as an individual entity which is dependent upon an internal harmony of all processes and a reaction tending towards the maintenance and the procreation of the individual. The problem of life lies hidden in a definite mutual relation and harmonious cooperation of physico-chemical processes (including those of the "environment") and still unknown processes which cannot be described in terms of the material processes which physics and chemistry have revealed to us.

To put the problem in this way means, in spite of its negative character a step forward, but at the same time it brings us to a point where it is more difficult than ever to tackle the problem further. For, the only starting-point we have is the unverifiable conviction that the characteristic of life must consist of unknown "linkings", processes, or "laws", which in any case are different from those which can be found in inorganic processes or are revealed by an analysis of the separate chains of causation in an organism.

This creates the great difficulty that we can no longer form a mental picture of the problematic entities and hence cannot render them intelligible.

At the moment we can only say that life cannot be interpreted in the anthropomorphic concepts with which we are familiar. Pictures based on our surroundings or ourselves turn out to be valueless when required to help us to understand what life is; up to now every picture, every model, every product of our imagination has failed when we tried to use it to explain life. Certainly: "Pictorial imagination is a wondrous blessing". "But", continues Alexander, "here is the provokingness of it. Just when we reach ultimate problems or ultimate conceptions, it deserts us. Either it is replaced by intellectual imagination or thought construction; — or it leaves us a prey to error or helplessness, which is often the case of the generality".

Similarly, it seems quite impossible to make a visual picture of an electron which bears any resemblance to well-known phenomena in our daily surroundings. The structure of the atom of modern theoretical physics can only be expressed in mathematical equations and abstract formulae, purely mental conceptions. The mental image of a pellet, subsequently a planetary system, which was the starting-point, has been destroyed. What is left of it is something like permanent elementary physical units, whose value is constant. As Dingle rightly remarks: "The scientist is no longer restricted in his thinking to elements which can be imagined and pictured, to elements which are clothed in the characteristics of phenomena. He may use any elements at all, provided that he can define them rationally, and he may suppose them to interact with one another in any conceivable way, provided that he can represent their behaviour by rational statements. This is legitimate, because the aim of science is to give a *rational* correlation of phenomena, but not necessarily an *imaginable* one".

### LIVING AND NON-LIVING SYSTEMS

So far, no consideration has been given to the important question, what is the difference between living and non-living systems?

As regards this point, it will be immediately observed that the characteristic of coordinated unity, which in the previous section was so much emphasized as a property of an organism, is not confined to a living system. In a crystal the particles of matter are also "organized" into a coordinated whole, which is characterized by a particular internal equilibrium and a typical constitution. Something like a "self-determining activity" is also a characteristic of an atom which in an ionized state "tries" to fill up its insufficiently occupied orbits by "catching" electrons. Moreover, an internal equilibrium is not only typical of physiological processes in an

organism, but just as much of e.g. the solar system. Nevertheless, many
will be inclined to believe that it is not difficult to define life and matter
as two contrasting entities. What can be more different than, for example,
a butterfly and a crystal, or a human being and a meteorite? The resemblance between them, in so far as it exists, that both can come into being,
mature, and perish, is only a very superficial one. For a crystal can be
formed from scattered particles which were before grouped together in
another way; while an organism always arises from another organism of
the same type of organization. The crystal grows through the accumulation of particles in accordance with a well-defined mathematical law; an
organism grows through metabolism. When a crystal disintegrates, it
resolves into parts which can in certain circumstances be re-grouped again
into the original form of the crystal. But when an organism falls into
decay, there is no possibility, not even in principle or in theory, of gathering
the inorganic and organic parts together again so as to form an individual
whole once more, except of course when in their scattered state they
are absorbed by and into other already existing and functioning living
organisms. Add to these other typical properties of organisms, such as a
progressive differentiation during the development of the individual, the
spontaneity of behaviour of the organism as a whole, birth, maturity,
and death — and it seems clear that a living system is founded on a
distinct guiding principle that differs fundamentally from that in a non-
living system.

There are, however, instances which could, as it were, be called border-
line cases between living and non-living. In some organisms we meet
with phases of rest, during which the metabolism, as far as we can ascertain, has practically come to a standstill. No one knows, for example, how
a Tardigrada can remain in a dried up condition of latent animation for
ten or more years, and then, when placed in water, swell out again and go
on living quite happily. Birth and death are not necessary attributes of a
living being, for one-celled creatures multiply by dividing up, and both
parts can, if we may put it thus, have an eternal life. Again, growing old
is a phenomenon which disappears when a tissue is removed from the
influences which proceed from the living whole, as Carrel showed by
separating tissue from the heart of a chicken embryo and culti-
vating it; since 1912 it has continued to grow, but it does not grow old.
Microbiology gives us another border-line case in the study of ultra-
filterable virus and bacteriophage types (fig. 123). Not only are the dimensions of e.g. the nucleoprotein virus sometimes smaller than those of a
large albumen molecule like haemocyanin (a non-living system), but on

the one hand it is capable of increasing to many times its original size, while on the other hand it can be precipitated into crystals, and then again when dissolved display renewed activity. Is this then a living system, or only a ferment or an autocatalyst, speeding up chemical reactions, which only actively increases when in contact with the living tissue of certain plants? Conversely, we know how in the atom, too, events occur which give an impression of spontaneity, or at least are not determinate.

With matters standing thus the question naturally arises, what minimum criteria a system must satisfy in order to be called a living system. Some scientists have attempted to define the criterion of life as freedom from the second main law of thermo-dynamics. This will be considered first. Then we shall discuss some of the theories which are directly connected with the development of modern atomic physics. It will be found that even negative results can be of importance.

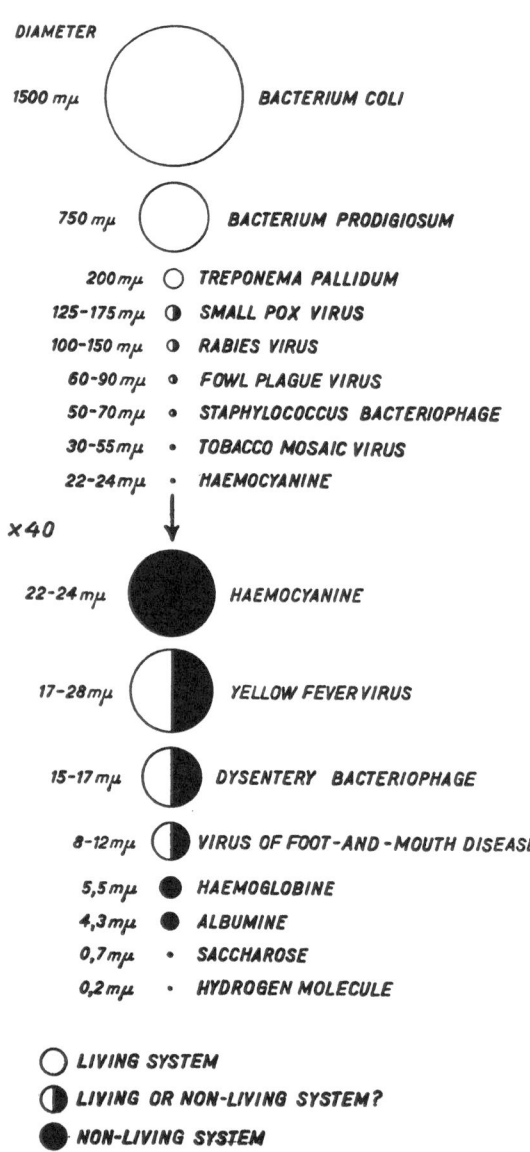

DIAMETER

1500 mμ — BACTERIUM COLI

750 mμ — BACTERIUM PRODIGIOSUM

200 mμ — TREPONEMA PALLIDUM
125-175 mμ — SMALL POX VIRUS
100-150 mμ — RABIES VIRUS
60-90 mμ — FOWL PLAGUE VIRUS
50-70 mμ — STAPHYLOCOCCUS BACTERIOPHAGE
30-55 mμ — TOBACCO MOSAIC VIRUS
22-24 mμ — HAEMOCYANINE

×40

22-24 mμ — HAEMOCYANINE

17-28 mμ — YELLOW FEVER VIRUS

15-17 mμ — DYSENTERY BACTERIOPHAGE

8-12 mμ — VIRUS OF FOOT-AND-MOUTH DISEASE
5,5 mμ — HAEMOGLOBINE
4,3 mμ — ALBUMINE
0,7 mμ — SACCHAROSE
0,2 mμ — HYDROGEN MOLECULE

○ LIVING SYSTEM
◐ LIVING OR NON-LIVING SYSTEM?
● NON-LIVING SYSTEM

Fig. 123. Comparison of dimensions of a few bacteria, virus, bacteriophage, and molecules. The upper series is 12,500 times enlarged, the lower series 500,000 times. (After Kluyver).

*Life and entropy.*

In his attempt to discover what the fundamental characteristic of life is, Ritchie (following Jeans) tells a story of imaginary intelligent beings who descended upon the earth from an outer world on which life had never existed. He invests these fantastic beings with a knowledge of mathematics, and of physical and chemical, but not biological, processes. In his opinion they would be particularly struck by the presence of accumulations of certain substances, such as they had not met with in their own world. They would be amazed at the enormous deposits of calcareous ooze, stretching over large areas of the bottom of the oceans, which, although sea-water has on the average only 0.012% of calcium carbonate, contain calcium carbonate in a proportion up to as much as 90%. They would be equally amazed at such formations as coral reefs, the chalk and limestone layers on the continents, the accumulations of carbon in vegetable tissues derived from the carbon dioxide in the atmosphere, the radiolarian and diatom oozes in which organisms have accumulated the very minute quantities of soluble silica in the ocean (never more than 1.5 parts per million) into some twelve millions of square miles of siliceous deposits, and the iodine in seaweed which has been extracted from the sea-water with a ten thousand-fold concentration.

These and other similar examples make us realize that organisms are capable of reversing the normal process of continual dispersal of earthly matter and turning it into the opposite, i.e. assortment and aggregation.

The question whether this concentrating power is typical of living organisms alone must, however, be answered in the negative. We need only call to mind the aggregations arising in the form of ores and exhalation products, or caused by alluvial agglomeration (e.g. gold nuggets) or by the differentiation of earthly matter into nucleus and encircling shells, or in magmas (by chemical processes acting in combination with the influence of gravitation), and so forth.

What ought to astonish Ritchie's non-terrestrial visitors is not the phenomenon of concentration itself, but only the fact that many concentrations are caused by living organisms. But Ritchie wanted to show by his examples of concentration caused by organisms that a living system is capable of going counter to the law of the increase of randomness by disintegration; for the second law of thermo-dynamics — also called the law of Carnot-Clausius — states that on the whole natural processes tend irreversibly towards a uniform level, a process called the increase of entropy.

Already in 1880 Maxwell asked whether in a living system something happens that is not in consonance with the general trend towards ever-increasing entropy which is found in a thermo-dynamic process [1]).

The principle of irreversibility involved in the law of Carnot-Clausius, was called anti-chance by Eddington. According to du Nouy "the study of life and evolution forces us to recognize that its action is logically required and has apparently always manifested itself in a "forbidden", ascensional direction to finally end in the thought and conscience of man". Burgers discussed this difficult problem at great length in 1943. Although he admits that these processes reveal normal statistical laws, and that they are subject to an increase of entropy, yet he thinks it probable that beside them, as the characteristic of life, a number of processes will appear of such kind that the statistical probability of the results will be upset, and it is doubtful whether the doctrine of entropy as such will then still be of any value. In any case his conclusion is: "It is impossible to avoid introducing non-physical elements when seeking to place that which typifies life and the urges which find expression in it".

According to von Bertalanffy classical thermo-dynamics is a fragmentary doctrine which is only applicable to closed systems. The organism as a whole, however, does not comprise a closed system. The characteristic state of the living organism is that of an open system, there is a constant import and export and therefore change of the components.

"Entropy must increase in all irreversible processes. Therefore, the change in entropy in a closed system must always be positive. But in an open system, and especially in a living organism, not only is there entropy production owing to irreversible processes but the organism feeds, to use an expression of Schrödinger's, from negative entropy, importing complex organic molecules, using their energy and rendering back the simpler end products to the environment. Thus, living systems, maintaining themselves in a steady state by the importation of materials rich in free energy, can avoid the increase of entropy which cannot be averted in closed systems".

The generalisation of thermo-dynamics and its extension to open systems

---

[1]) In strong contrast with the opinion of most authors is the theory of Blum. It is difficult to follow him in his remarkable conclusion that the principle of irreversibility involved in the second law of thermo-dynamics ,,supplies the necessary irreversibility which has been shown to be required for the evolutionary process, the direction of development being such as would always be accompanied by an increase of entropy, the return over the same pathway being prohibited by the fact that it would involve a decrease in entropy" (H. F. Blum, A considera-tion of evolution from a thermo-dynamic viewpoint — The American Naturalist, 69, 1935, p. 357).

was carried through, recently, by Prigogine. It leads to fundamentally new principles and seems to account for many characteristics of living systems that have appeared to be in contradiction with the laws of physics.

Moreover "the consideration of organisms as open systems yields quantitative laws of important biological phenomena".

*Life and Quantum-mechanics.*

In his treatise Burgers also points to the nature of the processes in the world of atomic dimensions. The fact that the processes of atomic dimensions are not strictly determinate has tempted many a scientist into drawing a parallel between the spontaneity in the behaviour of an organism and the "spontaneity", i.e. the element of indeterminacy, in quantum-mechanics. The events within an atom cannot be described on the lines of strict determinacy. Statistical laws of "probability" appear to reign instead. It cannot be foretold what situation will follow on the one observed. According to the calculations of the quantum-theory, a whole series of developments are possible, and it is impossible to ascertain which of the many possibilities will in fact take place. This is the point at which it becomes impossible to comprehend what happens in a physical law. Is this an elementary characteristic which serves to typify the phenomenon of life?

A phenomenon which may perhaps bear some relation to the indeterminacy of a quantum-jump is the sudden incidence of a saltation in the evolution of an organism. This seems possible, perhaps even probable, since in the genes (hereditary factors) we have to do with activities on an almost atomic scale. But even if the spontaneity in the incidence of a new saltation could be explained on these lines (it will be remembered that saltations can also be caused artificially, e.g. in Drosophila by treatment with X-rays), no light is thereby shed on the organic whole, the coordinatedness of a living organism.

P. Jordan has developed a theory, according to which the elementary events in the atom which do not proceed according to the principle of cause to effect, will be amplified by the structure of the cell and thus become manifest in perceptible phenomena. He regards life as "ein Wirken aus der Akausalität der Underwelt heraus in die kausal gebundene Oberwelt hinein". Apart from the fact that "non-causal" (which would erroneously suggest processes which involve caprice) does not accurately describe the indeterminacy of atomic processes, it must be remarked that the causality of "macroscopic" processes may not be contrasted with what happens in atomic processes as something entirely different, seeing that causality means nothing else than an event on such a scale that the uncertainty principle of Heisenberg may be ignored.

In their criticisms of Jordan's theory Hartmann, Büning and others point out that an amplifying effect, already long known in biology (e.g. in hormones and enzymes), may not be compared to the effect supposed by Jordan, seeing that it can still be defined within the laws of classical physics.

The laws governing a living system are different from those of a non-living system and cannot be described adequately and completely in terms of a physico-chemical theory. One of the consequences is that catalytic and autocatalytic processes — the importance of which was stressed by Jordan as well as by Jerome Alexander in his recent book *Life, its Nature and Origin* — display more pronounced effects in living systems as contrasted with non-living systems [1]).

Nevertheless, Jordan's amplification theory must not be rejected as wholly without value. Before long this ground will perhaps be approached from another side. Everyone knows, for example, how it is (sometimes) possible to make someone look up or turn his head or awake from sleep by staring hard at him (or her). How this is effected is, as far as I am aware, not known. But it is legitimate to suppose that there is an action which, originating from the central nervous system, causes a reaction in the person stared at which possibly arises from a quantum-jump, which, when "amplified", manifests itself in a sudden change in behaviour on the part of the person stared at. If something of this sort takes place (and perhaps it will one day be possible to treat this phenomenon experimentally and analyse it) yet the following points must not be forgotten. First, the behaviour of the person stared at is the reverse of what could be called spontaneous and free. Secondly, the effect is manifested in a living organism. Tracing a change in a (biological) behaviour to processes in the field of quantum or wave-mechanics may be compared to what happens when a saltation occurs as a result of treating the genes of a Drosophila with X-rays. None of these explanations penetrates to the secret spring of life itself.

Thus Büning wrote regarding P. Jordan's theory that the criterion of life is the amplification of "spontaneous" quantum-jumps:

"One who has spent any length of time in experimental study of living beings will not find an answer to his ultimate biological problem in the knowledge of amplifying life-centres, but in the study of the mutual

---

[1]) See in this connection also the publications by F. Pasquier and Mrs. Destouches-Février.

interaction of all parts, any one of which by itself has little of life in it, but which all together, by their mutual interaction, make life possible: we ought not to speak of living and dead parts, 'living molecules' [1]) and such, but we should say with the physiologist H. J. Jordan: 'The problem of all biology is the organization, or (as J. S. Haldane repeatedly showed) the coordination".

Often it seems to be forgotten that the indeterminacy in an atom takes place no less in a non-living system than in a living organism. Without further explanation it is not clear why and in what way these processes should in a living creature be so different from the indeterminacy of the processes in lifeless matter that the criterion of life is revealed thereby.

If the characteristics of a living system consist of a concourse of processes of intra-atomic factors the laws controlling these processes must be different from those governing non-living systems. Both the directive control so conspicuous in living systems as well as the way on which they interact with physico-chemical factors are still unknown.

Perhaps a first approximation of this difficult problem was that given by Schrödinger. In a condensed form the material carrier of the fundamental properties of life is found in the most essential part of a living cell, the chromosomes. These small and very complicated objects consist of a great number of still much smaller bodies — the genes — which are the bearers of definite hereditary factors. According to an estimate made by Delbrück a gene contains only about a thousand atoms. Even if there were a million that number of elementary particles would be much too small "to entail an orderly and lawful behaviour according to statistical physics — and that means according to physics. It is too small, even if all these atoms played the same role as they do in a gas or in a drop of liquid. And the gene is most certainly not just a homogeneous drop of liquid. It is probably a large protein molecule, in which every atom, every radical, every heterocyclic ring plays an individual role, more or less different from that played by any of the other similar atoms, radicals, or rings".

I should like to stress Schrödinger's conclusion that "we must be prepared to find it (a living system) working in a manner that cannot be reduced to the ordinary laws of physics".

The characteristic and determining entity in living beings and evolutive

---

[1]) That is what P. Jordan calls certain kinds of virus.

phenomena is different from the laws of chemistry and physics, in short of non-living systems.

It is a difficult but not a hopeless task for the future to unravel the determining and dominating vital factors in living systems. The greatest difficulty consists in the continuous interaction of the living system with the physico-chemical processes in the organism as well as with those of its environment.

If it sounds discouraging to some people that we cannot conceive or visualize the entity or principle governing a living system let them realize that no more can we visualize the factors or "laws" which a non-living system obeys.

Considering the problems of living beings and evolutive processes there is no need for a scientist to jump at once at the hypothesis of an irrational influence, at any rate no more than he would in considering the problems of non-living systems.

*The concept of complementarity.*

The negative result of the diverse attempts which have so far been made to define the characteristic of life as accurately as possible is nevertheless in itself not without importance. First it appeared that a life-principle, working as a directing force apart from matter, is an untenable conception. At the same time it became clear that a living system is bound up with material processes with which it forms a whole. Next it must be admitted that in living systems there are differences of degree as regards their "nature" or "intensity". Some are so entirely different from a non-living system that there is no difficulty in drawing a dividing line between them, while in other cases this seems almost impossible. And finally the characteristic of a living organism must be sought in unknown "linkings", which must therefore be taken to govern the coordination of the whole harmony embodied in the living creature.

According to Whitehead there is nothing else but events in a state of development. We cannot say that a system is at any moment in a clearly definable or certain condition, because the fundamental constituents of that system at any given moment cannot be determined. When, however, an array of elementary events remains constant for a certain length of time, we call what we observe "matter".

Objects are therefore to Whitehead "characters of events", occurrences which maintain a certain pattern of arrangement, like a river which keeps its shape, but in reality consists of an ever changing stream of particles; or like a reflection which we see of ourselves in water, which has a clear delineation, though that which composes it is continually altering.

As has already been remarked (pp. 182, 183), the important part of Whitehead's theory, to sum up the gist of it "libérée des tares de l'anthropomorphisme" [1]), is that he tries to comprehend life by means of an elementary process (spontaneity), which, however, is inherent in the whole of nature, so that we cannot expect in every case to draw a clear dividing line between life and matter. The processes pertaining to a non-living system are said to be distinguishable by the almost complete repression of this spontaneous element, so that a continuous repetition of the same pattern becomes characteristic of them. Where, however, this spontaneity, as he supposes, manifests itself in a living organism with greater intensity, he has to confess that this can only be ascribed to a sort of "linking" or "grouping", the origin and laws of which we do not know. Do not *yet* know, one should rather say, confidently looking towards the possible results of scientific research in the future.

Whitehead's synthesis is also interesting in that it reconciles the two opposites which are contained in the contrast between life and matter.

In this respect Bohr's concept of complementarity is also important, inasmuch as he has not hesitated to suggest that this concept is perhaps also applicable outside the strict confines of atomic physics, the original field in which he introduced it. In 1929 this Danish physicist offered the suggestion that physico-chemical processes on the one hand and the whole harmonious coordination that we call typical life on the other, are to be regarded as complementary phenomena. "The existence of life must be considered as an elementary fact that cannot be explained, but must be taken as a starting-point in biology, as in a similar way the quantum of action, which appears as an irrational element from the point of view of mechanical physics, taken together with the existence of elementary particles, forms the foundation of atomic physics" [2]). From which again it follows that a definition of life which only regards the physico-chemical aspect is one-sided and incomplete.

One of the consequences of the complementarity concept is that a theory concerning a living system ought to be of the type named subjectivistic in the sense of Destouches [3]). And from this Mrs. Destouches-

---

[1]) This expression is used (with reference to another problem) by de Broglie, 1941, p. 94.

[2]) Bohr, 1933, p. 458.

[3]) ,,Dans le cas où le système ne peut être observé qu'au moyen d'un appareil de mesure inéliminable même en hypothèse, et jouant un rôle essentiel dans l'expression des résultats des mesures, c'est-à-dire dans le cas où le système est microscopique, radicalement inaccessible aux sens, on montre que ce système ne comporte pas de grandeur d'état; toutes ses grandeurs ne

Février, in her paper on theoretical considerations in biology, concludes: "Ainsi une théorie biologique est nécessairement une théorie subjectiviste, ce qui en fixe déjà la structure générale et entraîne de nombreuses consequences; en particulier elle est essentiellement indéterministe ce qui introduit dans ce domaine un point de vue nouveau excluant à la fois mécanisme et finalisme absolues". Starting from quite different viewpoints we have arrived at the same conclusion (p. 197).

It is the task of the science of the future to seize every opportunity that occurs of investigating the complementary aspect of life. After all that has been said it is clear that this is the most difficult problem which mankind, since the remotest ages of its civilisation, and science, more recently within somewhat better defined limits, have ever had to face. Furthermore, it must always be remembered — to quote the words of Bohr — "that the problem of distinguishing between living and non-living is perhaps one which cannot be comprehended in the ordinary sense of the word".

### UNITY AND HARMONY

In the midst of the many changing opinions, relative values, and unpicturable products of thought of modern science, the question automatically arises with more urgency than ever before: what has man still left to cling to in his scientific researches, what beacon to guide him on his way, or in other words: what is the standard of valuation in science?

As the first axiom Newton wrote in his *Principia*: "We are to admit no more causes of natural things than such as are both true and sufficient to explain their appearances", and he continued: "Philosophers say that Nature does nothing in vain, and more is in vain when less will serve, for Nature is pleased with simplicity and effects not the pomp of superfluous causes".

From human behaviour in society to the formulae of abstract science, everywhere the old adage *simplex veri sigillum* is held in high regard.

We abhor complicated situations, and we have a comfortable feeling when we have succeeded in reducing something complicated to simple terms. But it would be wrong to call it simply a "feeling". The principle of simplicity has been accepted since Newton's time not because it seems

---

sont pas simultanément mesurables et les résultats des mesures effectuées sur lui ne sont pas des propriétés intrinsèques, mais expriment les propriétés du complexe observateur-appareil-système. On a alors une théorie subjectiviste, ainsi nommée parce qu'il est impossible, même en droit, de séparer les connaissances que l'on acquiert sur le système des moyens employés pour les obtenir". (Destouches-Février, Theoria Vol. XV, 1949, pp. 81/82).

to be self-evident, as it was in the Middle Ages and even in the time of Copernicus and Galileo, but because observation shows us that it holds universally.

Now it will perhaps be remarked that the modern idea of the structure of the atom is definitely less simple than the colourless pellet which formed its historical starting-point. Further, that space and time in the "classical" sense could be called obvious realities, which as such were comparativily easier to grasp than the concepts to which the theory of relativity and quantum-mechanics have led us. Next, that the earlier theories about light, both that of Newton and that of Huyghens, are in principle less complicated than the modern views of wave-mechanics concerning the photon. That at one time the origin of folded mountain-chains was simply explained by the old theory of contraction as a sort of shrinking comparable to the rumpled peel of a drying fruit, while now we have to refer to a number of complicated processes, to many of which a large query must be attached. And so we could go on for some time.

Among the newer theories there are, therefore, many which are undeniably more complicated than their predecessors. Nevertheless, in all these cases we discard the old explanation and adhere to the new theory, or only even to the tentative suggestion, in spite of its having a more complex character. Not merely because the original conception was found to be useless and a new, if necessary more complicated, theory had therefore to be accepted on account of its practical usefulness; but rather because the new theory reveals something of a hitherto unknown and unsuspected harmony, a correlation between a number of phenomena which formerly could not but be thought to be heterogeneous and unconnected with one another. It is in this harmonious embrace, which interconnects what at first seemed heterogeneous, that the simplicity, the *veri sigillum* just mentioned, is to be found. In short, the *only* reason which induces us to replace a theory by a new one is that the latter offers *a rational correlation of more factors of our experience and knowledge than its predecessor.* Five examples may first be given to explain this further.

In the history of our planet certain phenomena of a very heterogeneous nature have come to light; considerable upward and downward movements of the sea-level took place over long periods of time with a rhythm which attracts attention by reason of its monotonous regularity, large continental ice-caps grew and melted again, basins and troughs appeared as dents in the outer covering of the earth, mountain-chains towered on high and were eroded away, granitic and other "plutonic" rocks originated in the crust of the earth during certain periods. One day it becomes possible to see these heterogeneous phenomena in their relation to one another. There is found to be a definite mutual connection. They are no longer loose strains of music; they appear to be woven together as in one grand fugue. Thus we can see the history of the earth before us as the score of a symphony. And we rejoice at this concord of harmonious sounds on account of the unity which enfolds the different parts. The simile of the *symphony of the earth* is preferable to the easier picture of a shrivelling apple, because it gives us a more-embracing harmony.

Seldom has a hypothesis been propounded which is so difficult to understand as the quantum-mechanics of Planck and Einstein. It is a doctrine which is the least likely to appeal to the imagination. Yet it is attractive by reason of the harmonious connection which it gives to hitherto very heterogeneous phenomena. It is accepted as a further approximation of truth, even though no one can say what the meaning is of the fundamental h, Planck's constant, which turns up everywhere where a natural phenomenon is probed to its depths, as if it were an incomprehensible ground-tone of the universe.

When Bohr succeeded in bringing the theory of the quanta, the atom-model of Rutherford, and the formula for spectral lines of Balmer Rydberg nd Ritz together in one view with his atomic theory of the year 1913, he presented us with a by no means simpler model of the atom; and we know that the science of the atoms has since then had to pass on to still less simple constructions. But the possibility of uniting heterogeneous phenomena in one view gives us so strong a confidence that we are on the right track, that here again we prefer the composite and unrepresentable to the original naive conception with which we began.

This same confidence, or property of our mode of thought, is also clearly manisfested in our dislike of the concept of dualism. Physical science was faced with an intolerable deadlock when some of the phenomena of light could only be explained by means of Newton's corpuscular theory, and others only by the wave theory of Huyghens and Fresnel. A feeling of rest after strife is given by the resolving of this dual nature of matter and radiation in de Broglie's wave-mechanics.

Biologists who thought that the nineteenth century mechanical age of science had been vanquished found themselves compelled to regard living and non-living systems as two different principles, each with its own autonomy. But such a dualism cannot give satisfaction; and so we are glad that Bohr's concept of complementarity offers a chance of removing this dualism by absorbing it into the harmonious unity of a monistic conception of a higher order.

In all these examples we have to do with a principle which amounts to a striving to reduce as many phenomena as possible to one common denominator. It is obvious that this striving is in reality actuated by nothing else than a desire for simplicity.

Moreover, this principle of simplicity undeniably contains an aesthetic element. In searching for the connection between and the meaning of the observed phenomena we are often largely guided by an appreciation of harmonious beauty.

To mediaeval thinkers spheres and circles seemed to be the most perfect figures. It was the principle of simplicity with which they were mainly concerned.

It was, for example, chiefly aesthetic considerations which led Copernicus to his new view of the universe. We need further only recall Kepler's *Harmonia mundi*, and Plato, whose theory of ideas is well known and whose harmony of the spheres has become an immortal allegory. And is it chance that the Greek word "Kosmos" means not only the world, but also ornament and orderliness?

Homo sapiens also forms a part of this supreme ornament, this orderliness and unity which is the characteristic of the Kosmos [1]). As stated already on page 181, Whitehead takes as his starting-point what we experience in ourselves. The same tendencies are then promoted to become a basic principle for the understanding of all nature. But man is thereby himself made a part of the nature which surrounds him. Whitehead's conclusion is therefore that in the Kosmos there is no absolute independent reality, but that all things are closely bound up with one another, and that Homo sapiens is a part and a link in the whole development of the Kosmos: "The universe is always one, since there is no surveying it except from an actual entity which unifies it".

The whole universe is not governed by a reality, conceived as lying outside it; it is in all things permeated by the reality which is the universe itself in its entirety.

It has already been remarked that the appreciation of what is true and that of what is beautiful in a scientific theory are frequently closely interwoven. Some are of opinion that this is only a sort of mediaeval survival in our mode of thought. Others attach to it a deeper significance, which is not the outcome of the predilections of a particular age. In one of his recent works de Broglie writes: "A doctrine which succeeds in bringing about at one stroke a huge synthesis, by showing the underlying analogy

---

[1]) Many parallels to these ideas may be found in the doctrine of the ancient Upanishads. Think only of the main theme of the old Brahmanism: Atman = Brahman.

between phenomena which superficially have no relation to one another, undeniably produces in the mind of the thinker an impression of beauty, and persuades him to believe that it contains a great deal of truth".

This brings us to a problem which might easily lead us far beyond the proper bounds of natural science. Although the temptation to do so must be resisted, it is nevertheless desirable to devote a few words to this aspect of the problem. For here there is opened up the possibility of a synthesis which comprises fields of mental activity which until recently seemed to most physicists to be separated by an impassable gulf.

We remember how Newton succeeded in bringing physics, especially mechanics, into one view with a large part of astronomy; how Woehler caused the dividing line between organic and inorganic chemistry to fade; how the modern atomic theory has merged chemistry and physics into one; how Bohr united the undesirable dualism of life and matter in a compound of opposites as it were; while finally Whitehead included even the problems of man himself, as we have seen, in one all-comprehending synthesis of the universe. Such a synthesis does not mean that it is possible to find a path which leads directly from, say, the understanding of the motions of a celestial body or the structure of the nucleus of an atom to the regions of, say, ethics and art. But it does mean that even these can be seen as different facets of one and the same reality. Many a scientist of to-day arrives anew at an opinion which has frequently been uttered before in the last twenty centuries of our civilisation (beginning with Plato and the Neo-Platonists like Plotinus), albeit in different forms according to time and circumstances. For a long time science and philosophy went their own too widely separated ways, and sometimes it was only a matter of chance that the one came to know of the results arrived at by the other. For example, it was only pure chance that one of America's greatest sons, Benjamin Franklin, became acquainted in London with the writings of Shaftesbury, whose stimulating optimism deeply moved him and had a great influence on his conception of the world. What Shaftesbury revealed to him was the gift of seeing in the Kosmos the aesthetic unity and harmony of the whole. The same idea, in principle, has been expressed in other words by a modern physicist like de Broglie.

It goes without saying that a scientist of the twentieth century cannot possibly be a blind follower of Plato, Plotinus or Shaftesbury. Even these authors themselves, if resurrected, would undoubtedly find their manuscripts out of date and needing to be entirely reworked into a new edition adapted to the present state of knowledge. In their modernized treatises they would presumably stress the common aspects not only of diverse branches of the sciences of nature, but also of the fields of reason (logic) and moral values (ethics), of psychology and of art. For in these different departments of mental activity, rational and non-rational, man is guided by one and the same urge to correlate experiences which, when viewed only superficially, are quite distinct. It is his constant longing to reveal a connection and a unity which belongs to the realm of metaphysical aesthetics. But let us dwell no longer on this fascinating subject. For a discussion of the harmonious synthesis of everything that can be known and perceived, thought and believed, would certainly go far beyond the scope of this chapter, which is concerned only with a few aspects of life and its evolution.

Only this one point of view, the gist of the foregoing remarks, need finally be brought forward. The farther science tries to penetrate into the secrets of the universe, the less is human imagination able to grasp the results attained. These results are indeed in ever increasing number out of the reach of mental pictures. On the other hand, however, the horizon of rational correlation is steadily becoming wider than was ever apparent before; the synthesis, though incomprehensible to our senses in its interwoven details, is ever growing through the unification of the numerous entities of which it seems to be composed. Hence, the disappointment of man's habitual longing for pictorial representation is amply compensated by a much greater and more sublime harmony; the whole of nature discloses a beautiful unity, which seems to extend from the remotest distances of the universe to the inmost depths of ourselves.

### REFERENCES

A symposium on evolution was arranged by the *Institut International des Sciences Théoriques* in Paris, Octobre 10–15th 1949. This chapter is adapted from the author's address to the participants of the Symposium delivered under the title: Aspects paléontologiques de l'évolution.

Additions have been worked up from two other lectures, one before the Society of Science (*Natuurkundig Genootschap*) in Groningen, March 1943, the other one before the *Studium Generale* at Delft, February 1949.

The theoretical part of the present chapter was translated by Mr. C. P. Hierneiss, B. A., B.C.L. (Oxon) from the Dutch manuscript which covers part of the 3d edition of *Leven en Materie* (Nijhoff, The Hague 1946) as well as parts of the 2d edition of *De Beeldenstorm der Wetenschap* (Nijhoff, The Hague 1945).

The following list of literature is strongly abbreviated from the bibliographies in these two books, while a few recent publications have been added.

ALEXANDER, J., *Life its nature and origin* (Reinhold, New York, 1948).
ALEXANDER, S., *Artistic creation and cosmic creation* (Proc. of the British Academy, 13, 1927).
BAAS BECKING, L. G. M. *Notes on the determined and the undetermined in Biology* (Acta biotheoretica, vol. VIII, part L/II, 1946, p. 18).

BAUR, E., *Einführung in die experimentelle Vererbungslehre* (Borntraeger, Berlin 1919).

BERGSON, H., *L'évolution créatrice* (Alcan, Paris, 1921).

BERTALANFFY, L. VON, *The theory of open systems in physics and biology* (Science, 111, 1950, pp. 23–29).

BOHR, N., *Light and Life* (Nature, 131, 1933).

BOHR, N. *La théorie atomique et la description des phénomènes* (Gauthier-Villars, Paris, 1932) p. 110.

BROGLIE, L. DE, *Continu et discontinu en physique moderne* (Albin Michel, Paris, 1941).

BROGLIE, L. DE, *L'Avenir de la science* (Plon, Paris 1941).

BROGLIE, L. DE, *Physique et microphysique* (Paris 1947).

BÜNING, E., *Quantummechanic und Biologie* (Die Naturwissenschaften, 31, 1943).

BURGERS, J. M., *Over de verhouding tusschen het entropiebegrip en de levensfuncties* (Verh. Nederl. Akad. van Wetensch. Amsterdam, 18, 1943).

BURGERS, J. M., *Het entropie begrip en de rol daarvan bij levensfuncties.* (Natuurk. Voordrachten, Diligentia, 1946).

CUÉNOT, L., *Théorie de la préadaptation* (Scientia 16, 1914).

CUÉNOT, L., *Invention et finalité en biologie* (Flammarion, Paris 1941).

DECUGIS, H., *Le vieillissement du monde vivant* (Masson, Paris 1941).

DESTOUCHES-FÉVRIER, P., *Considérations théoriques en biologie* (Comptes rendus des séances de l'Académie des Sciences, 225, 1947, pp. 466–468).

DINGLE, H., *Science and human experience* (Williams and Norgate, London 1931).

DOLLO, L., *Les lois de l'évolution* (Bull. Soc. Belge de Géologie, VII, 1983).

DRIESCH, H., *Philosophie des Organischen* (Engelmann, Leipzig, 1921).

GOLDSCHMIDT, R., *The material basis of Evolution* (New Haven 1940).

GOLDSCHMIDT, R. B., *Ecotype, Ecospecies and Macro-evolution.* (Experientia, vol. IV, fasc. 12, 1948. p. 465–472).

GOUDOT, P., *Sur la deuxième loi de la thermodynamique en biologie* (Comptes rendus des séances de l'Académie des Sciences, 225, 1948).

HUXLEY, J., *Essays of a biologist* (Chatto and Windus, London, 1923).

HUXLEY, J., *Evolution, a modern Synthesis* (Allen and Unwin, London 1944).

HUXLEY, J., *Man in the modern world* (London, Chatto and Windus, 1947).

JEPSEN, G. C., SIMPSON, G. G. and MAYER, E., *Genetics, Paleontology and Evolution. A symposium* (Princeton University Press, 1949).

JORDAN, H. J., *De causale verklaring van het leven* (Noordholl. Uitg. Mij, Amsterdam 1942).

KLUYVER, A. J., *'s Levens nevels* (Handel. 26e Natuur- en Geneesk. Congres, Utrecht 1937).

LILLIE, R. S., *The living and the non-living.* (The American naturalist 68, 1934).

LULL, R. S., *Organic evolution* (Macmillan 1921).

MATISSE, G., *Le Rameau vivant du monde. Le déchiffrement des faits* (Presses universitaires de France, 1947).

MATISSE, G., *Le Rameau vivant du monde, Philosophie biologique* (Presses universitaires de France, 1949).

MAYR, E., *Systematics and the origin of species,* (New Yotk, Columbia Univ. Press, 1942).

OPARIN, A. I., *The origin of Life* (Macmillan, New York 1938).

OSBORN, H. F., *Equidae of the Oligocene, Miocene, and Pliocene of North America* (Mem. Americ. Mus. Nat. Hist. IV, Ser. II, 1918).

OSBORN, H. F., *The Titanotheres of ancient Wyoming, Dakota and Nebraska* (Monogr. 44 of the U.S. Geol. Survey, Washington 1929.)

OSBORN, H. F., *Proboscidea* (Americ. Mus. Nat. Hist. I, 1936, II, 1942).

PASQUIER, F., *Autocatalyse des systèmes protéiques* (Comptes rendus des séances de l'Académie des Sciences, 224, 1947; 225, 1948).

PRIGOGINE, I., *Etude thermov-dynamique des systèmes irréversibles* (Paris, Durrod, 1947).

RITCHIE, J., *Perspectives in evolution* (Rep. British Association for the Advanc. of Science, 1939)

RENSCH, B., *Die Transspezifische Evolution* (Ferdinand Enke, Stuttgart 1947)

SIMPSON, G. G., *Tempo and Mode in Evolution* (Columbia Univ. Press., 1944).

SIMPSON, G. G., *The meaning of evolution* (Yale Univ. Press, 1949).

SIRKS, M. J., *De ontwikkeling der biologie* (Noorduyn 1942).

SCHINDEWOLF, O. H., *Palaeontologie, Entwicklungslehre und Genetik* (Bornträger, Berlin 1936).

SCHRÖDINGER, E., *What is Life? The Physical aspect of the living cell* (Cambridge Univ. Press 1945).

SWINNERTON, H. H., *Outlines of paleontology* (3d ed. Arnold, London 1947).

THOMPSON, D'ARCY W., *On Growth and Form* (Cambridge, Univ. Press, 1942).

TRUEMAN, A. E., *The use of Gryphaea in the correlation of the Lower Lias* (Geolog. Mag. LIX, 1922).

UMBGROVE, J. H. F., *Leven en Materie* (Martinus Nijhoff, 's-Gravenhage 3d edit., 1946).

UMBGROVE, J. H. F., *Evolution of reef corals in the East Indies, since Miocene time.* (Bull. of the Americ. Assoc. of Petroleum Geologists, 30. 1946, pp. 23–31).

VANDEL, A., *L'Homme et l'Evolution* (Paris, Gallimard, 1949).

WELLS, H. G., HUXLEY, J., and WELLS, G. P., *Science of Life* (Cassell 1931).

WHITEHEAD, A. N., *Science and the modern world* (1933).

WHITEHEAD, A. N., *Modes of Thought* (1938).

# INDEX